ADLARD COLES BOOK OF
DIESEL ENGINES

FIFTH EDITION

MEL BARTLETT

**ADLARD
COLES**

LONDON · OXFORD · NEW YORK · NEW DELHI · SYDNEY

ADLARD COLES
Bloomsbury Publishing Plc
50 Bedford Square, London, WC1B 3DP, UK

BLOOMSBURY, ADLARD COLES and
the Adlard Coles logo are trademarks of
Bloomsbury Publishing Plc

First published in 1998 in Great Britain as *The RYA
Book of Diesel Engines* by Adlard Coles Nautical

Reprinted 2000, 2002
Second edition 2002
Third edition 2005
Fourth edition 2011
This edition published 2018

A catalogue record for this book is available from the
British Library

Library of Congress Cataloguing-in-Publication data has
been applied for

ISBN: PB: 978-1-4729-5540-1;
eBook: 978-1-4729-5541-8;   ePDF: 978-1-4729-5542-5

2 4 6 8 10 9 7 5 3 1

Typeset in Optima and Futura by Sheer Design and
Typesetting.

Printed and bound in India by Replika Press Pvt. Ltd.

Bloomsbury Publishing Plc makes every effort to ensure
that the papers used in the manufacture of our books
are natural, recyclable products made from wood
grown in well-managed forests. Our manufacturing
processes conform to the environmental regulations of
the country of origin.

To find out more about our authors and books visit
www.bloomsbury.com and sign up for our newsletters

CONTENTS

FOREWORD TO THE FIRST EDITION

Every year the rescue statistics published by the RNLI show that the most common cause of lifeboat launches to pleasure craft is machinery failure. In the case of motor cruisers this does not come as any great surprise; one might expect loss of motive power to figure high in the list of problems. The fact that engine failure is also the most common cause of sailing cruiser rescues is less predictable and serves to confirm just how important it is to keep the engine in good running order.

In response to these statistics, the RYA introduced a one-day course on diesel engine operation. The syllabus is, very broadly, the material covered in this book, although the depth into which it is possible to go in such a short course is inevitably rather limited. The aim of both the course and of this book is not to create instant diesel mechanics, but to provide boatowners with a better understanding of how their engines work and what they must do to keep them working.

While it would be great if everyone could carry out all the servicing and repairs on their own engines, this is not a realistic proposition; few boatowners have the time to become skilled mechanics and not many boats carry the tools, spares and equipment to provide the full workshop support needed for complex repairs.

What is achievable by every owner is an understanding of the importance of routine engine management, how to rectify the most common and relatively simple problems which occur and how to recognise the warning signs that an engine needs expert attention.

Fortunately, most diesel engines are reliable and relatively trouble free in operation, so boatowners do not spend a high proportion of their time confronted by smoky exhausts, screeching temperature warning alarms or engines that obstinately refuse to start. Hence much of the knowledge acquired on a diesel engine course is seldom put into practice. This reinforces the need for a clear, comprehensive reference book, both to back up the knowledge gained on a course and to provide a guide for those who prefer to teach themselves.

Bill Anderson,
Former RYA Training Manager

FOREWORD TO THE FIFTH EDITION

For more than two centuries – ever since two French inventors built the world's first compression-ignition engine and demonstrated its potential by using it to propel a boat – manufacturers have been competing against each other to produce engines that are lighter, more powerful, quieter, more economical or more reliable.

Over the past 30 years or so, however, legislation has taken over from commercial pressure as the main driving force behind technological development.

In 2003, for instance, the European Recreational Craft Directive (RCD) was amended to include noise and emission limits. Yet these limits were relatively lax compared with those that were already being applied to road vehicles. As most pleasure craft engines are 'marinised' versions of car, truck or bus engines, it was inevitable that the advances in engine design that were required to meet the road vehicle regulations would eventually find their way on to boats.

In many respects, that's a good thing. The best way to reduce emissions is to burn less fuel, so – in general terms – a clean engine is likely to be smaller, lighter and cost less to run than its dirty predecessor. What's more, the advanced fuel injection systems required to make it cleaner are also inherently more reliable.

But some things haven't changed.

Twenty years ago, in his Foreword to the first edition of this book, Bill Anderson – former training manager at the Royal Yachting Association (RYA) – wrote that 'every year the most common cause of lifeboat launches to pleasure craft is machinery failure'.

It still is.

And for every lifeboat call-out, there are countless, less dramatic situations that are caused by some kind of mechanical failure: a day's racing missed because the engine wouldn't start, an embarrassing crunch in a marina because a control system has jammed, or precious days of a summer cruise wasted because the engine overheats …

You may not be able to strip and rebuild a modern engine with a screwdriver and an adjustable spanner like my grandfather did, but you can still go a long way towards avoiding these situations by giving your engine a bit of tender loving care – particularly making sure that its fuel is clean, and that it is cool and lubricated.

That's what this book is all about.

Mel Bartlett,
February 2018

1 WHY CHOOSE A DIESEL?

I still remember the time when, as a child, I was given my first ballpoint pen. It was one of those with a knob on top that, when pressed, made the nib emerge and when pressed again made it retract. Like most small children, I amused myself by clicking it in and out for a while. The clicking, I recall, seemed much more fun than writing.

It wasn't long, though, before that novelty wore off – and not much longer before my new pen had 'come to bits' as I tried to find out how it worked. I suppose most of us have done much the same thing, and I'm quite convinced that the outcome of that experience determines our future attitude to all things mechanical.

If you are one of those for whom the pen never clicked again, take heart. Remember that for all their apparent complexity, engines depend on a sequence of simple processes. They don't have souls, or wills of their own, so if you can make sure that those processes go on happening in the right order, your engine just has to keep on running. The flip side of the coin is that if you don't, your engine can't keep going out of any sense of affection, loyalty, or self-preservation!

That much, at least, applies to all engines, whether you're talking about the electric motor of a vacuum cleaner or the jet engines of an airliner. Every type of engine, however, has its own strengths and weaknesses that make it more suitable for some purposes than others. That's why you don't find jet-powered vacuum cleaners or electrically powered aircraft, and why you're more likely to have a diesel engine powering your boat than your lawnmower.

Compared with a petrol engine, for instance, a diesel engine is likely to be expensive, heavy and slow to respond. On most boats, though, these drawbacks are worth putting up with in order to take advantage of a diesel's main attributes:

♦ Reliability
♦ Long life expectancy
♦ Low running costs
♦ Non-explosive fuel

Even a diesel engine, however, will deteriorate if it is neglected, and could ultimately corrode away to become a useless lump of rusty metal. To take advantage of its reliability and long life expectancy it needs to be looked after. Of course you can pay someone else to do the work for you, but that eats away at the advantage of low running costs.

The aim of this book is to help you get the most out of the capital invested in your engine, by making the most of the advantages you've already paid for – reliability, longevity and economy.

A fringe benefit of doing your own maintenance will be familiarity with your engine and the tools you use to work on it. Then, if things *do* go wrong, you have a sporting chance of either being able to solve the problem yourself, or of giving a professional mechanic something more to go on than 'it just sort of stopped'.

THE BASIC ENGINE

2

As I pointed out in Chapter 1, diesel engines don't have souls or wills of their own, but depend on a sequence of simple processes.

The most fundamental of all those processes takes place deep inside the engine. It's the one that gives internal combustion engines their name, because it involves burning air and fuel inside a confined space.

THE BASIC PROCESS

The confined space is the *cylinder* – a vertical tube, machined into the heavy metal *block* that accounts for most of the engine's weight and bulk. The top of the cylinder is closed by another heavy casting called the *cylinder head*. Tunnels in the cylinder head allow air and exhaust gas to flow in or out of the cylinder, controlled by *valves*.

The bottom of the cylinder is formed by the *piston*, another machined metal casting that is designed to slide up and down inside the cylinder, with springy metal *piston rings* forming an almost gas-tight seal between the piston and the cylinder walls.

Don't bother, for the moment, about how we get a mixture of fuel and air to burn inside the cylinder: just accept that as it burns it produces a mixture of water vapour, carbon dioxide and small quantities of some more unpleasant gases, such as sulphur dioxide and oxides of nitrogen. It also gets very hot.

The rise in temperature makes this gaseous cocktail expand – increasing the pressure within the cylinder, and driving the piston downwards. The piston is attached to a connecting rod, or *'con rod'*, whose other end is coupled to the *crankshaft*. Just as the cranks of a bicycle convert vertical movements of the rider's legs to a rotary movement of the wheels, the crankshaft converts the downward thrust of the piston into a rotary movement of the shaft.

One end of the crankshaft carries a heavy metal *flywheel*. Once the flywheel has started turning, its momentum keeps it going, so the crankshaft keeps turning with it – pushing the piston back up the cylinder. As it does so, one of the valves in the cylinder head opens, allowing the hot gases to escape.

As soon as the piston reaches the top of its travel, the still-spinning flywheel and crankshaft drag it back down again. At this point, the exhaust valve shuts and the inlet valve opens, allowing fresh air to flood into the expanding space inside the cylinder.

This time, as the piston reaches the bottom of its stroke, the inlet valve closes. With both valves shut, and the momentum of the flywheel driving the piston back up again, the air inside the cylinder is compressed.

If you compress any gas, it gets hot. You can feel the effect for yourself by putting your finger over the outlet hole of a bicycle pump and pumping the handle. Even after several hard strokes, a bicycle pump is unlikely to develop

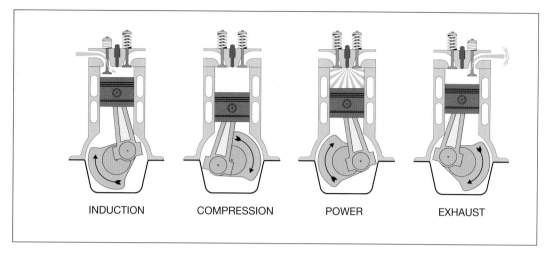

| INDUCTION | COMPRESSION | POWER | EXHAUST |

Fig 1 The four-stroke cycle.

more than about 100 psi, but the pressure inside a diesel engine's cylinder rises to over 500 psi in less than $1/100$ second. Its temperature rises, as a result, to something in the order of 800° C.

Diesel fuel doesn't burn easily under normal conditions, but if you spray a fine mist of it into hot pressurised air, it will ignite spontaneously. The engine's fuel system is designed to do exactly that – producing, in the cylinder, the burning mixture of air and fuel required to start the cycle all over again.

So there you have it: the basic operating cycle of a diesel engine, made up of four distinct strokes of the piston. You can think of them, if you like, as 'suck, squeeze, bang, blow', though in more conventional terminology they're called Induction, Compression, Power and Exhaust.

VALVES

The work of the valves is vital to the whole sequence: they have to open and close at precisely the right moments, allowing an unrestricted flow of air or exhaust gas when they're open, yet forming a perfectly gas-tight seal when they're shut.

Each valve is roughly mushroom-shaped, with a long straight stem and a flat circular head, whose edge is bevelled and precision-ground to match the slope of the hardened *valve seat* that surrounds the mouth of the tunnel in the cylinder head. For most of each cycle, each valve remains shut, pulled firmly against its seat by one or two very strong *valve springs*. It's opened, when necessary, by a component called a *rocker*, like a miniature seesaw that pivots on another shaft running across the cylinder head.

Meanwhile, a component called the *camshaft* is being driven by the crankshaft, but at half the crankshaft's speed. On it are carefully machined bulges, called *cams*, that are shaped and positioned so that each in turn pushes upwards against a rocker at the right moment in each cycle. As one end of a rocker is pushed upwards, the other end moves downwards to push the valve open.

Although the principle is standard, there are plenty of variations on the theme. The camshaft, for instance, may be driven by gears, or by a chain and sprocket system, or by a toothed rubber belt, and it may be mounted high on the engine with the cams pushing directly on the rockers; or lower down and relying on *push rods* to transmit the movement of the cams to the rockers. In this case, the ends of the push rods don't rest directly on the cams but sit in small bucket-shaped components called *tappets* or *cam followers*. In some engines, the cam followers are fitted with rollers to reduce wear. In others, they are designed to rotate so as to spread the wear more evenly, while some engines have hydraulic tappets which adjust themselves to correct for wear as it happens.

Whichever of these applies to your particular engine, do bear in mind that the whole system will have been set up so that each valve opens and closes at precisely the right moment in the cycle. Small amounts of wear and tear can be corrected by means of a simple adjustment, but it's asking for trouble to tinker with the gears, belt or chain unless you know exactly what you're doing.

THE TWO-STROKE CYCLE

It seems rather wasteful to have the piston going up and down like a yo-yo, but only producing power on one of its four strokes. There is an alternative, called the *two-stroke* cycle. Apart from the fact that it produces power on every second stroke of the piston, the diesel two-stroke has very little in common with its petrol-oil counterparts on lawnmowers and outboards, and its use is mainly confined to the very large engines that drive ships. The one exception is the Detroit Diesel range, which includes two-strokes down to 270 hp.

They are physically different from conventional four-stroke diesels in that they have no inlet valves. Instead, air is pumped into the cylinder by a mechanical blower – a *supercharger* – through *ports* halfway up the cylinder walls. (See Fig 3.)

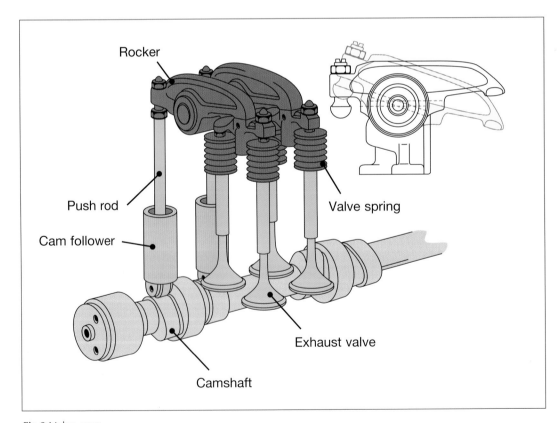

Rocker

Push rod

Cam follower

Valve spring

Exhaust valve

Camshaft

Fig 2 Valve gear.

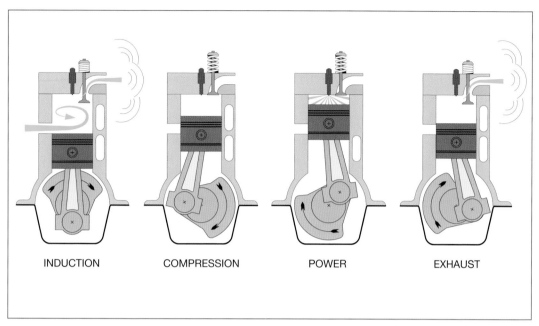

| INDUCTION | COMPRESSION | POWER | EXHAUST |

Fig 3 The two-stroke cycle.

When the piston is at the bottom of its travel, these ports are above the level of the piston, so, with the exhaust valve open, clean air flows into the cylinder and blows the previous stroke's exhaust gas out of the top.

As the piston rises, the exhaust valve shuts, and the piston itself closes the inlet ports, trapping the air inside the cylinder. The compression stroke continues, just as in a four-stroke engine, and is followed by the power stroke driving the piston downwards.

Just before the piston reaches the level of the inlet ports, however, the exhaust valve opens, allowing the exhaust gas to start escaping. As the piston descends still further, it uncovers the inlet port, allowing fresh air into the cylinder, to start the sequence all over again.

The advantages and disadvantages of two- and four-stroke engines are pretty evenly balanced: power for power, two-strokes are smaller and lighter, but are slightly less fuel-efficient, and because they are produced in very much smaller numbers they tend to be relatively expensive. Most of their repair and maintenance procedures are similar, though, so we'll concentrate on the more common four-stroke engine throughout this book.

VARIATIONS ON A THEME

One apparently subtle variation is the distinction between *direct* and *indirect* injection.

Fig 1 illustrating the four-stroke cycle shows a direct injection engine: the fuel is sprayed directly into the cylinder. In practice, the top of the piston is usually carved away to form a hollow, called the *combustion chamber*, shaped to ensure that the fuel and air mix as thoroughly as possible.

In an indirect injection engine, the piston crown is usually flat, and the combustion chamber is deeply recessed into the cylinder head, with only a narrow opening between it and the cylinder. The idea is that the turbulence created when air from the cylinder is forced into the combustion chamber ensures more thorough mixing of the air and fuel, and a more progressive increase in cylinder pressure during the power stroke.

Historically, at least, indirect injection engines have been regarded as quieter and cleaner but harder to start, because the cylinder head absorbs a lot of the heat created during compression. Unfortunately, the heat lost to the cylinder head and the effort required to force air and burning gas in and out of the combustion chamber make them rather less fuel-efficient overall.

Developments in piston design are now allowing modern direct injection engines to catch up with the indirect engine's advantages without the drawbacks, so indirect injection seems destined to fade away.

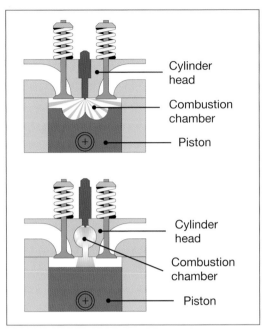

Fig 4 Direct injection (above) and indirect injection.

... THINGS TO DO

CHECKING VALVE/ROCKER CLEARANCES – ONCE PER SEASON

There isn't much an amateur mechanic with a limited tool kit can (or should) do to the major components inside the engine apart from making sure that it has a good supply of fuel and air, and clean lubricating oil.

You can, however, check and adjust the gap between the rocker and the valve. There has to be a gap – usually about the thickness of a fingernail – to allow for the different rates at which the various components expand and contract as they warm up. Without it, there's a very real risk that the valves won't shut completely: they may even come into catastrophic contact with the pistons. If the gap is too large, the valves may not open as far as they should, and the engine will certainly be noisier than it should be.

a. Read the engine manual to find out what the valve/rocker clearances should be, and whether they should be adjusted with the engine cold or at normal running temperature. Note that the clearances for inlet valves may be different to those for exhaust valves, because exhaust valves get hotter.

b. Remove the rocker cover – a relatively thin metal box on top of the engine, usually with the oil filler cap in the middle. Some engines have a separate rocker cover for each cylinder, or for each of two or three groups of cylinders.

c. Check the gap on each valve in turn, when the valve is completely closed and the gap is at its widest. There are two ways of finding out when this happens. On a multi-cylinder engine, the best way is to find the 'magic number' for your engine by adding one to the number of cylinders. For a four-cylinder engine, for instance, the magic number is five.

d. Turn the engine slowly by hand, if necessary using a spanner on the crankshaft (big nut on the lowest of the pulleys at the front of the engine). Watch the rockers moving as you do so, until the two rockers for one cylinder are 'on the rock' – that is, when one is rising and the other falling – signifying that this particular cylinder is at the end of its exhaust stroke and just beginning its induction stroke. Subtract the number of this cylinder from the 'magic number' to find the number of the cylinder that is ready to have its valve clearances checked. If, for instance, you have a four-cylinder engine and number 2 cylinder's valves are on the rock, number 3 cylinder is ready, because $5 - 2 = 3$.

e. On a single-cylinder engine, the clearance for one valve should be checked when the other valve is fully depressed. You can use this approach for a multi-cylinder engine, but it will take longer!

f. Slacken the lock-nut on the rocker whose clearance you are about to adjust, and then unscrew the threaded adjuster about one or two turns.

g. Set a feeler gauge to the clearance specified in the engine manual, and slip it between the valve stem and the rocker. Gently wiggle the feeler gauge while tightening the adjusting screw, until you can feel the feeler gauge being nipped between the valve stem and the rocker.

h. Leave the feeler gauge in place, and hold the adjusting screw with a screwdriver while you tighten the lock-nut. When it's tight, wiggle the feeler gauge again to check that you haven't upset the adjustment: you should feel a slight resistance, but it shouldn't be jammed tight.

i. Repeat the process for each valve in turn, then replace the rocker cover, making sure that the cork or rubber sealing gasket is smooth, undamaged and properly seated.

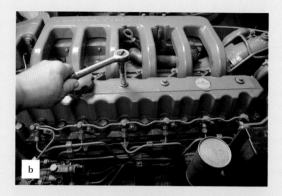

b

e

g

FUEL SYSTEM

3

Otto Diesel's original patent application for what we now know as a diesel engine was pretty vague about the kind of fuel it might use: he even suggested coal dust as a possibility. Some boatowners seem almost equally vague: every year lifeboats have to tow in boats that have simply run out of fuel!

There's more to getting fuel into an engine, however, than simply pouring the stuff into the tank.

Diesel fuel doesn't burn very easily, and in order to burn quickly, cleanly and reliably it has to be in the form of fine droplets, like an aerosol spray. You'll remember from the previous chapter that the air in a diesel's cylinders is made hot by being compressed to 20 or 30 times its normal atmospheric pressure, so producing an aerosol spray inside the cylinders means that the fuel has to be at an even higher pressure – in the order of 2,500 psi.

It's also essential for the proportions of fuel and air to be exactly right, so each squirt of fuel has to be very accurately measured. If you think of a typical four-cylinder diesel developing 80 hp when it's running flat out at 4,000 rpm, it will be burning about 4 gallons of fuel an hour. Each cylinder will be receiving 2,000 squirts of fuel every minute – making 8,000 squirts per minute, or 480,000 squirts per hour. Each squirt, then, must be less than 10 millionths of a gallon, 0.04 ml, or less than a hundredth of a teaspoon. At low loads the amount of fuel sent to the cylinders has to be even less.

It's hardly surprising, then, that the fuel system includes some of the most sophisticated and expensive parts of the engine, responsible for achieving pressures of almost 200 atmospheres, measuring doses of fuel accurate to less than a thousandth of a millilitre, and repeating the process perhaps half a million times an hour!

THE BASIC SYSTEM

The fuel system starts, however, with the crudest component of all: the tank. It's worth bearing in mind that a full tank can be very heavy, so it needs to be well supported and secured against the boat's motion. A big tank – anything over about 5–10 gallons – should include internal baffles to stop the fuel sloshing about, and any tank needs a vent, or 'breather', to let air in as the fuel is used up.

Unfortunately, the fuel received from the hose may not be perfectly clean, and the air that comes in through the breather will almost certainly be moist enough to allow condensation to form inside the tank. The end result is that the tank will include some dirt and water.

To prevent this reaching the engine, the engine installation should include a component known as a *primary filter, pre-filter, separator, sedimenter* or *filter-agglomerator*, usually mounted on a bulkhead in the engine compartment rather than on the engine itself.

The *lift pump* is responsible for pulling the fuel out of the tank, through the primary filter,

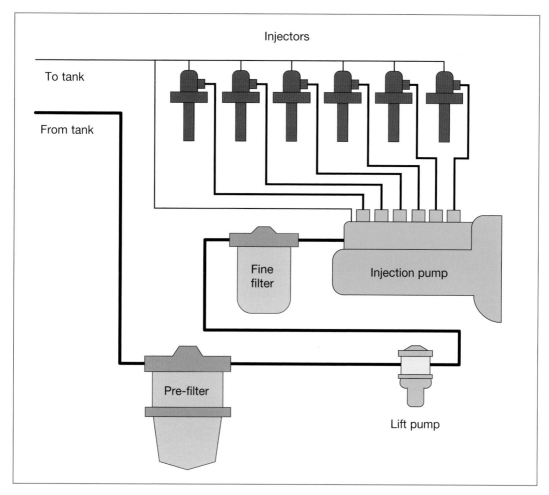

Fig 5 General layout of fuel system.

and passing it on to the rest of the system. In most cases, it's a simple diaphragm pump, very much like a miniature version of a manual bilge pump. It's driven by the engine, but usually has a hand-operated *priming lever* so that you can pump fuel through the system without running the engine.

The fuel then passes through another filter, sometimes known as the *main filter* or *secondary filter* or *fine filter*, whose job is to remove particles of dirt that – at less than a thousandth of a millimetre in diameter – may be too small to see, but that are still capable of wearing the very finely engineered surfaces of the rest of the system.

If a diesel engine has a 'heart', it has to be the *injection pump*, because this is where the fuel is measured and pressurised.

Injector pipes, with very thick walls to withstand the pressure, carry the highly pressurised fuel from the injection pumps to the *injectors* that spray it into the cylinders.

Some of the fuel that is pumped to the injectors, however, never actually reaches the cylinder but is returned to the tank through a *leak-off pipe*, or *return line*.

THE SINGLE-ELEMENT INJECTION PUMP

There are three main types of injection pump, of which the simplest is the kind found on single-cylinder engines. Even if you have a multi-cylinder engine, it's worth knowing a bit about the *single-element* 'jerk' pump, because many multi-cylinder engines use derivatives of it.

The principle is much like that of a bicycle pump or an old-fashioned bilge pump, with a piston (usually called the *plunger*) moving up and down inside a cylinder. A hole called the *spill port* in the side of the cylinder allows fuel to flow into the cylinder when the plunger is at the bottom of its travel. As the plunger rises, however, it covers the port to shut off the flow and trap some fuel in the cylinder. As it continues to rise, the trapped fuel has to go somewhere, so it escapes by lifting the delivery valve off its seat, and flowing out into the injector pipe.

PRINCIPLE OF THE JERK PUMP

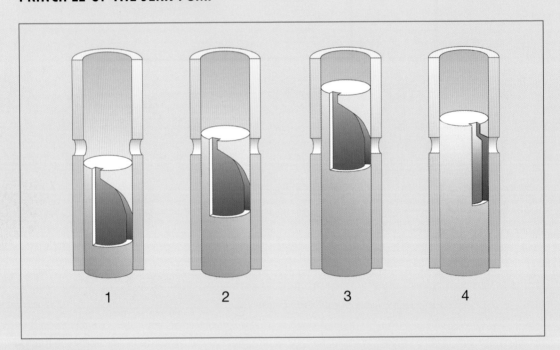

1   2   3   4

Fig 6 Jerk pump.

1. When the plunger is at the bottom of its travel, fuel flows into the pump cylinder through one of the ports.

2. As the plunger rises, it blocks off the ports and pressurises the fuel, driving it out of the top of the pump cylinder.

3. As the piston rises further, the helical cut-out reaches the spill port: fuel can flow down the groove and out through the spill port. The pressure is released so no more fuel reaches the injector.

4. Rotating the plunger means that the cut-out reaches the spill port at an earlier stage in the plunger's travel. The effective stroke of the plunger is shortened, so less fuel is delivered.

IN-LINE INJECTION PUMP

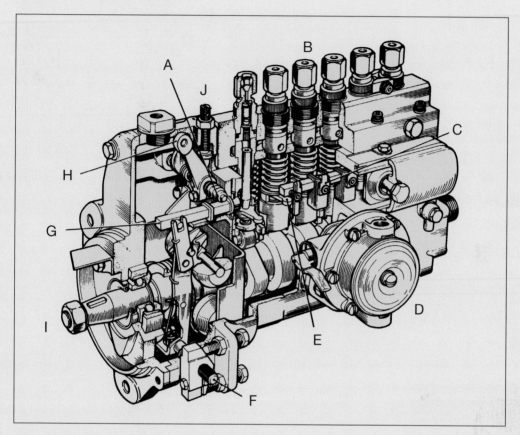

Fig 7 An in-line fuel injection pump.

KEY

A. Excess fuel button for cold starting.

B. High pressure fuel line connectors that feed the injectors. Six in this case for a six-cylinder engine.

C. Control fork that moves levers on the plunger arm on each pump to control the quantity of fuel injected.

D. This model has the low pressure fuel pump built on to the side of the injection pump. This is a diaphragm type driven from the injection pump's camshaft rather than from the main engine camshaft.

E. The actuating arm that along with C moves the pump element to control the amount needed for injection at various engine speeds.

F. Control lever connected by cable to the helm position.

G. Control rod assembly, which is moved by F and a combination of the excess fuel device, the engine governor and the stop control to provide exactly the right control of the pumping elements to suit the particular running or stopping conditions.

H. Stop lever.

I. Cam and roller cam followers, which drive the pumping elements. This is a pump that requires the gallery to be topped up with engine oil for the internal lubrication of the moving parts.

J. Maximum fuel stop screw, usually has a seal placed through it to prevent tampering.

The measuring part of the fuel pump's job is taken care of by a spiral-shaped cut-out in the side of the plunger. As the piston nears the top of its travel, the spiral cut-out eventually comes level with the spill port in the side of the cylinder, allowing fuel to flow round the spiral and out of the spill port.

Pushing or pulling on a toothed rod called the *rack* makes the plunger rotate, so the spiral can be made to uncover the spill port at any stage in the plunger's stroke, varying the amount of fuel that is delivered without having to change the distance the plunger actually moves.

This is significant, because the up-and-down movement of the plunger is achieved by the action of a cam, very similar to the cams that operate the valves in the engine's cylinder head.

It's worth noting that thin metal packing pieces called *shims* are usually fitted between the base of the pump and the cylinder block or crankcase. Increasing the number of shims raises the pump body, so the ports are higher, which means that the pump doesn't start delivering fuel until slightly later in the cycle. In other words, the number and thickness of the shims has a critical effect on *timing* – the moment at which fuel is sprayed into the cylinder – so if you remove the fuel pump for any reason, it's essential to make sure that you retain all the shims and put them back when the pump is re-installed.

THE IN-LINE INJECTION PUMP

A few multi-cylinder engines use a separate single-element fuel pump for each cylinder, but it's more common to find all the separate elements combined into a single component that looks rather like a miniature engine. It's called an *in-line pump* because it consists of several jerk pumps in line, driven by a camshaft in the pump body instead of in the engine block.

THE ROTARY INJECTION PUMP

The *rotary* or *DPA injection pump* is lighter, more compact, and can cope with higher engine speeds than the in-line type, so it's eminently suitable for small, high-revving engines. Unfortunately, it's also more vulnerable to dirty or contaminated fuel and – unlike an in-line pump that may fail on one or two cylinders but keep going on the others – a DPA pump that goes wrong will often pack up altogether.

The reason for this 'all-or-nothing' operation is that a DPA pump consists of a single *high pressure pump*, distributing fuel to each injector in turn through a spinning rotor.

The lift pump supplies fuel to the injection pump at one end, where a vane-type *transfer pump* – similar in principle to the engine's raw-water pump – increases its pressure. The fuel then flows to the high pressure pump through the *metering valve*, which controls the amount of fuel that will be delivered to the engine's cylinders.

The high pressure pump consists of two small plungers built into a rotor. Fuel from the metering valve flows into the space between the two plungers, forcing them to move apart. As the rotor turns, however, bulges on the *cam ring* that surrounds it force the plungers back inwards.

Fuel, now at very high pressure, is driven out of the space between the plungers and through a drilling in the rotor, which directs it to each injector pipe in turn.

INJECTORS

The *injectors* convert the tiny squirts of high pressure fuel into an atomised spray in the cylinders. They are usually cylindrical in shape, about 6 in (15 cm) long and 1 in (25 mm) in diameter, but are clamped into the cylinder head so that only a couple of inches of the injector body and a couple of pipe connections are visible.

ROTARY INJECTION PUMP

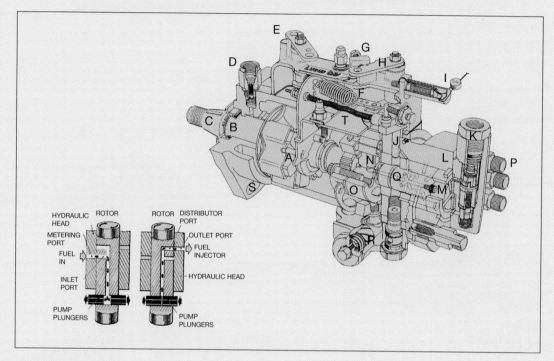

Fig 8 Rotary injection pump.

KEY

A. Centrifugal governor weights provide sensitive speed control.

B. Front bearing oil seal and retaining circlip.

C. Tapered drive shaft.

D. Back leak connection feeds excess fuel that has also helped lubrication of the pump back to the fuel filter.

E. Shut off lever, hand operated by cable control.

F. Return spring to hold speed control lever against idle stop.

G. Idling speed control stop.

H. Speed control lever usually connected to helm position by cable control system.

I. Maximum speed stop and adjusting screw sealed to prevent tampering.

J. Fuel metering valve, governor controlled.

K. Low pressure fuel inlet with nylon filter below it.

L. The stationary hydraulic head, which houses the transfer pump (M) and the distributor rotor (Q).

M. The transfer pump, which transfers low pressure fuel from inlet (M) to high pressure plungers (N) via metering valve (J).

N. High pressure pump plungers are driven outwards by fuel pressure from (N) and pushed inwards by the lobes on the cam ring (O).

O. Cam ring.

P. High pressure outlet pipe connections to injectors.

Q. The distributing part of the rotor contains a central axial passage (dotted) and two radially drilled ports. The distributing port aligns successively with each high pressure outlet port to P, there being one for each cylinder of the engine. A similar number of inlet ports in the rotor align successively with a single port in the head, called the metering port, and admits the fuel from (M) under the control of the governor. See inset.

R. Fully automatic advance device.

S. Pump fixing and locating bolt slot that allows rotation of pump about axis for timing. Score marks across engine and pump flange can help re-install pump to same timing position.

T. Governor spring.

THE PRINCIPLE OF A MECHANICAL GOVERNOR

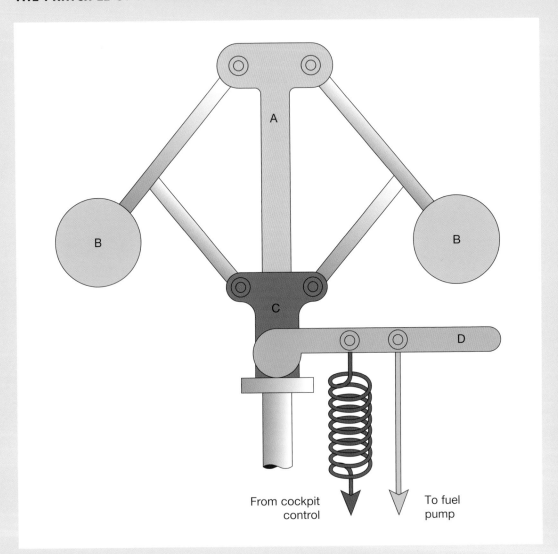

Fig 8a The principle of the mechanical governor.

The shaft (A) is driven by the engine, so as the engine speed increases, the weights (B) try to fly outwards. The linkage (C and D) is arranged so that this movement tells the fuel pump to slow the engine down.

The cockpit control is connected to the spring. When the control is pushed forwards, for higher engine speeds, the increasing tension in the spring makes it more difficult for the flywheel weights to slow the engine down, so the engine speed increases.

The balance between the governor weights and the spring tension keeps the engine running at a constant speed, set by the cockpit control, even if the load varies.

The mechanical governor inside a diesel injection pump is more sophisticated than this, but the principle is identical.

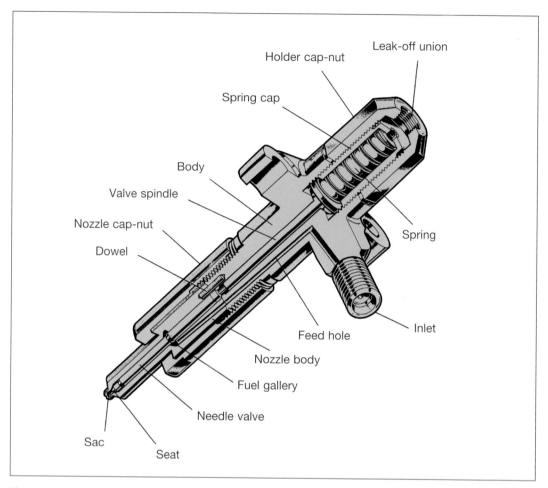

Leak-off union

Holder cap-nut

Spring cap

Body

Valve spindle

Nozzle cap-nut

Dowel

Spring

Feed hole

Inlet

Nozzle body

Fuel gallery

Needle valve

Sac

Seat

Fig 9 Injector.

The injector body is basically a tube, almost completely filled by a *needle valve, push rod*, and a strong *spring*. Fuel from the injection pump enters the side of the injector from the injector pipe, and then flows down a narrow passage to the *pressure chamber*, just above the nozzle.

The *nozzle* is sealed off by the needle valve, which is held in place by the push rod and spring. When the injector pump delivers one of its pulses of fuel, the pressure within the pressure chamber rises sufficiently to lift the needle valve off its seat. Fuel then rushes out of the nozzle so quickly that it breaks up into a spray. Of course, this sudden escape of fuel means that the

pressure in the pressure chamber drops again, allowing the needle valve to snap back into its seat to stop the flow.

Although the movements of the needle valve are very small, they happen so quickly that lubrication is essential. This is achieved by allowing some of the fuel from the pressure chamber to flow up the injector, past the needle valve and push rod, and out through the leak-off pipe at the top to return to the tank.

If too much fuel took this route, it would entirely defeat the object of the exercise: the pressure in the pressure chamber would never rise enough to lift the needle valve, so no fuel would get into the cylinder. The fact that it doesn't is entirely due to the very high precision engineering of the injector, which keeps the clearance between the needle valve and the injector body down to something in the order of 0.001 mm (about 40 millionths of an inch). That's so small that if you were to strip an injector and leave the body on the bench while you held the needle valve in your hand, your body heat would expand the needle valve enough to stop it going back into its hole!

There are three reasons for mentioning this, of which the first is to make the point that you should never strip an injector: it may look rugged, but it's so finely engineered that injector servicing is definitely a job for a specialist company. The second reason is that it goes a long way towards explaining why new injectors can cost several hundred pounds each, and the third is that it explains why all those filters are so important: the tiniest specks of dirt can be sufficient to abrade the surface of the needle valve enough to increase the leak-off to such an extent that the injector doesn't open properly, or to wedge the valve open and allow fuel to drip out of the nozzle instead of forming a fine spray.

The same applies to injection pumps, because there is nothing an amateur mechanic can achieve by tinkering with them, other than a lot of damage. Even the apparently simple job of removing an injection pump is more complicated than it may seem, because re-fitting it involves adjusting it to make sure that the squirts of fuel are delivered to the right cylinders at the right time: it needs confidence and the right workshop manual.

wanted cleaner, quieter cars and lorries, and legislators wanted to be seen to be doing something.

So, in December 1987, the European Union introduced Directive 88/77/EEC. Mirroring similar legislation imposed by other countries, including the USA, it was the first of a series of directives and regulations that imposed increasingly stringent limits on the exhaust emissions from road vehicles.

The effects are dramatic: it would take a convoy of 60 trucks built to the current (2018) 'Euro V1' regulations to produce the same amount of particulates ('soot') as a single one of the same power 25 years ago.

The rules also demand massive reductions in the amount of carbon monoxide (down by 87 per cent from 1988), hydrocarbons (down by 95 per cent) and oxides of nitrogen (down by 97 per cent).

None of this could be achieved by minor refinements: something far more significant was required. That 'something' was a major re-think of fuel systems to produce a finer spray from the injector nozzle by using higher pressures, and to give better control over the timing and the amount of fuel injected by using electronic control systems.

Europe-wide emission regulations for recreational craft were not introduced until 2003, and are still less demanding than the rules applied to road vehicles. But so many marine diesels are based on engines designed for cars, trucks and buses that anyone with a new-ish diesel should be aware that their engine is likely to have an *electronic control system* and either *unit injectors* or a *common rail* injection system.

HIGH-TECH FUEL SYSTEMS

The last few years of the twentieth century saw growing concerns, worldwide, about the use of fossil fuels and atmospheric pollution. Customers

ELECTRONIC CONTROL

A key part of any conventional fuel pump is the governor. At its simplest, this consists of a set of weights connected to the shaft of the pump. As the engine speed increases, the pump shaft turns faster, so the weights try to fly outwards. As they do so, they operate a mechanical linkage that reduces the amount of fuel being sent to the injectors. This, of course, slows the engine down, allowing the governor weights to move inwards again.

The engine control, in the cockpit or wheelhouse, is connected to the governor by a spring. By adjusting the engine controls the helmsman adjusts the spring tension so as to increase or decrease the speed at which the shaft has to turn before the weights move outwards far enough to slow the engine down.

The aim of all this is partly to stop the engine over-revving, but it also means that when you – the user – set the throttle for a particular engine speed, the governor will keep the engine running at that speed even if the loading varies.

Simple mechanical governors like this have been used to control machinery for centuries: you can see rudimentary versions in watermills, windmills, and on steam engines. But their place is now being taken by electronic control units (ECUs) – sometimes called electronic control modules (ECMs).

As well as monitoring the throttle position and shaft speed, an ECU is connected to a variety of sensors around the engine, monitoring factors such as the air temperature and pressure, the fuel temperature and pressure, and the coolant temperature. Taking all these factors into account, it can then tweak the timing and duration of each squirt, and even break a single squirt into several separate squirts to give the most efficient combustion.

The downside of electronic control systems is that they have done away with the rugged simplicity that used to be one of the great advantages of a marine diesel: a modern diesel is just as dependent on electricity as a petrol engine – just as vulnerable to the effects of salt,

water and corrosion – and can potentially be brought to a standstill by the failure of a single sensor. Overall, modern diesels are still more reliable than their older counterparts – but only if they are kept clean and meticulously serviced.

UNIT INJECTORS

Unit injectors, in principle, are almost a retrograde step: they take us back to the days when each injector had its own high pressure pump. In a unit injector, however, there is no separate high pressure injection pump, as such. Instead, each injector includes its jerk pump element (see page 10) built into the top of the injector.

A unit injector incorporates a high pressure jerk pump inside the body of the injector itself.

Fuel flows from the tank, through a pre-filter and then – driven by a low pressure lift pump – through a fine filter, just as in a conventional fuel system (Fig 5). From there, still at low pressure, the fuel flows directly to the injectors.

The jerk pump mounted on top of each injector pressurises the fuel to about 2500 bar (>35,000 psi) and dispenses the correct quantity of fuel to the injector nozzle.

In some cases, the pump is mechanically driven. Each cylinder has three rockers instead of the usual two. Two of the three rockers open the valves, exactly as they do in a conventional engine, while the third one operates the pumping element of the injector.

This alone brings potential benefits, because it does away with the need for injector pipes, and allows the injector to operate at far higher pressures.

But the real advantage comes when the unit injector is hydraulically operated and electronically controlled.

In the current generation of unit injectors, a hydraulic pump provides a source of high pressure engine oil to open the needle valve in the injector. The timing of each squirt and the amount of fuel injected is controlled by an electrical solenoid in response to signals from the engine's electronic control unit.

COMMON RAIL INJECTION

Like unit injector systems, the key feature of *common rail* or *reservoir* fuel systems is that the metering and control functions have been taken away from the injector pump altogether.

A low pressure lift pump pulls fuel from the tank through a pre-filter, and pushes it through a fine filter and on to the high pressure pump, just as in the conventional fuel system shown in Fig 5. But in a common rail system, the pump's

sole job is to provide a more-or-less constant supply of fuel at enormously high pressure – up to 3000 bar (43,000 psi) – to a tubular reservoir called the 'rail' mounted alongside the cylinders. From there, the pressurised fuel flows to the injectors.

The needle valve inside each injector is opened and closed by a solenoid in the head of the injector, triggered by electrical pulses from the ECU.

Because the high pressure pump is pumping continuously, rather than in measured squirts, it can maintain an almost constant pressure in the rail and injector pipes, so the pressure at the injector doesn't fluctuate as it does in a conventional system. This produces a more consistent spray pattern.

The high pressure also means that the droplets of fuel in the spray are smaller, so they burn more efficiently.

And because the injector valve opening is controlled electrically, rather than by surges of pressure, the timing can be varied, and it can even be made to open and close several times in a single cycle.

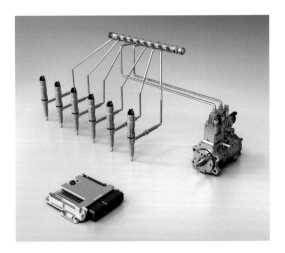

The main components of a common rail fuel system. Bottom: the ECU; right: the high pressure pump; top: the common rail; left: the injectors.

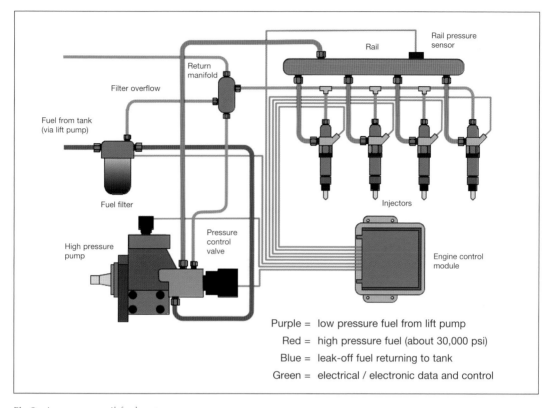

Purple = low pressure fuel from lift pump
Red = high pressure fuel (about 30,000 psi)
Blue = leak-off fuel returning to tank
Green = electrical / electronic data and control

Fig 9a A common rail fuel system.

... THINGS TO DO

There is absolutely nothing an amateur mechanic can or should do to the internal working parts of an electronic control unit, to a unit injector, or to a common rail fuel system, without specialist expertise and equipment. Meddling with any of them is likely to do more harm than good, and may be dangerous.

Never try to 'bleed' a common rail system by 'cracking' (loosening) any of the joints in the high pressure pipes between the high pressure pump and the rail or between the rail and the injectors.

♦ Regular checking and changing of fuel filters and water traps is more important than ever.

♦ Visually inspect electrical connections, and clean/tighten if necessary.

♦ On rocker-driven unit injectors, check and adjust the rocker clearances in accordance with the manufacturer's instructions and the procedure outlined on pages 6–7.

SAFETY FIRST

Diesel fuel can cause skin problems, especially in people who have become sensitised by repeated contact. Avoid the risk by using protective gloves and by keeping your hands clean.

The fuel leaving the injection pump is at such high pressure that it can penetrate skin. This is particularly true of the very fine droplets that leave an injector at high speed. Never expose yourself to high pressure diesel.

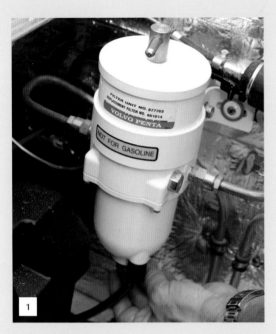

Draining the pre-filter.

1. DRAINING THE PRE-FILTER

The pre-filter is the part most likely to be affected by water or dirt from the fuel tank, so it should be checked frequently. The optimum interval will vary widely, depending on how clean your fuel is to start with, and how quickly you're using it, as well as on the filter itself, but after every ten hours' running is usually about right.

Many pre-filters have a transparent bowl at the bottom, so you can see any dirt or water at a glance. If yours doesn't have this, or if you can see a layer of dirt or water collecting at the bottom, you will need to drain it.

a. Slacken the drain screw at the bottom and allow the contents of the filter to run off into a suitable container such as a jam jar until clean fuel emerges.

b. Shut the drain screw, being careful to avoid using excessive force (it's hollow, and can snap easily), and then dispose of the contaminated fuel carefully.

c. Some pre-filters have a replaceable element similar to that in a cartridge-type fine filter, and which should be replaced in much the same way.

2. REPLACING THE FINE FILTER

The fuel filter should be changed at least once a season, or after about 200 hours' use. Start by cleaning the area around the filter, and placing a bowl or rags underneath to catch any spills. If your filter is below the level of the fuel in the tank, shut the fuel cock on the tank, but remember to open it again before attempting to start the engine. In any case, you will have to *bleed* the system before starting the engine.

SPIN-ON FILTERS

a. Use a strap or chain wrench to unscrew the filter canister. If this isn't available or doesn't work, try a large pair of gas pliers or a set of stillsons (pipe wrench).

b. Smear the sealing ring with a thin film of fresh oil, then spin the filter on until the sealing ring just touches the filter head.

c. Tighten the filter another half turn by hand. Do not over-tighten it by using any kind of tool.

Replacing a spin-on filter.

CARTRIDGE FILTERS

a. Unscrew the central bolt to release the filter body.

b. Remove the cartridge, and replace it, making sure that the various springs and washers are replaced in the correct order, and that the filter is the right way up. Make sure the old rubber sealing ring isn't stuck to the filter head, and replace it with the new one supplied with the filter.

c. Replace the complete assembly, making sure the filter body is correctly seated, and tighten the retaining bolt.

WATER TRAP FILTERS

Some filters have a bowl designed to trap water underneath the filter cartridge.

The sequence of photos below shows the fitting of a new cartridge:

a. Slacken the drain tap in the bowl and drain off the contents of the filter. Then unscrew the bolt that protrudes from the centre of the bowl.

Fitting a new cartridge to a water trap filter.

b. Reassemble the filter with a new cartridge and the new seals that are supplied with it – noticing that the upper and lower seals are different.

c. Tighten the central bolt gently and carefully, applying no more than about 10 lb to the end of a typical spanner.

3. BLEEDING THE FUEL SYSTEM

Even a very small amount of air in the fuel system can be enough to stop a conventional diesel, because if air bubbles reach the injector pipes they can act as shock absorbers that prevent the pressure from rising sufficiently to open the injector's needle valve.

If the engine suddenly stops or misfires, or if you have let air into the system by running low on fuel or changing a filter, you may have to remove the air by 'bleeding' the system.

Special hollow bolts called bleed screws are provided for the purpose.

In principle, the process involves working from the tank towards the engine, slackening each bleed screw in turn until clear diesel comes out, then tightening that screw and moving on to the next one.

Look for bleed screws at the highest point on top of each filter and on the injector pump. If you can't find a bleed screw, it is usually enough to slacken one of the pipe unions instead, but bear in mind that many newer engines are 'self-bleeding': they will bleed themselves if you crank the engine over for long enough.

Never try to 'bleed' a common rail system by 'cracking' (loosening) any of the joints in the high pressure pipes between the high pressure pump and the rail or between the rail and the injectors.

Bleeding the fuel system.

AIR SYSTEM

4

Fuel, by itself, is of no use whatsoever: it needs oxygen from the air outside in order to burn. At the most basic level, this happens of its own accord: as the piston falls during the induction stroke, air rushes in past the open inlet valve to fill the expanding space. Then, when the compression and power strokes are complete, the exhaust valve opens and the rising piston pushes the exhaust gas out ready for a fresh charge of clean air.

In practice, though, the engine needs an *air filter* to stop dirt, moisture and bits of rubbish being sucked into its cylinders, and it needs an *exhaust system* to dispose of the hot exhaust gases safely and quietly. To save having a separate filter and exhaust pipe for each cylinder of a multi-cylinder engine, the incoming air is fed to the cylinders through a tubular structure called the *inlet manifold*, and the exhaust gases are carried away through a similar structure called the *exhaust manifold*.

AIR FILTERS

Unlike their cousins that power tractors and earth-moving machinery, marine diesels usually operate in a relatively clean environment: there's little danger of them having to contend with straw, dust or roadside litter. This means that their air filters can be relatively simple, so some engines operate perfectly well for years with little more than a metal box with a few baffles in it.

Most, however, feature something a little more sophisticated, involving either wire gauze or porous paper.

Paper tends to restrict the air flow, so to make up for this its area has to be increased by being folded into a concertina shape. It's also difficult to clean, so once a paper filter becomes clogged it has to be replaced with a new one.

Wire gauze doesn't restrict the air flow as much, but it is less effective because the gaps between the strands of wire are bigger than those between the fibres of paper. To counter this problem – and to minimise corrosion – wire gauze filters need to be dipped in oil from time to time, so that dust sticks to them instead of passing straight through.

EXHAUST SYSTEMS

When it comes to exhaust systems, the boot is on the other foot: road vehicles and agricultural machinery have an easy time of it. Their engines are in compartments that are open to the atmosphere but sealed away from their drivers and passengers, so all that's required is a pipe connected to the exhaust manifold, with a few baffles to reduce the noise. A few marine installations adopt a similar 'dry' exhaust system, usually in the form of an exhaust pipe sticking straight up from the engine compartment, with a weighted flap to stop rain or spray running down inside and heat resistant lagging to minimise the risk of fire or burns.

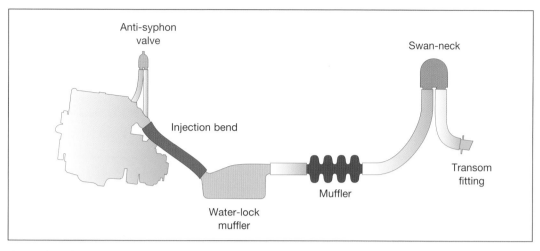

Fig 10 The exhaust system.

For pleasure craft, though, *'wet'* exhausts are pretty well standard, with water from the engine's cooling system used to cool the exhaust gas. The water is mixed with the exhaust gas in the *injection bend*, where it almost immediately turns into steam but in doing so reduces the temperature of the exhaust gases from almost 500° C to about 70° C – cool enough to allow flexible tubing and GRP to be used for the rest of the exhaust system.

At that reduced temperature, the steam condenses back into water. That is why the mixing takes place in a bend: it protects the engine against the possibility of the cooling water running back through the system and into the cylinders.

If the engine is below the waterline, or very close to it, however, the injection bend alone is not enough: there's a danger that water already in the exhaust might set up a siphon effect that would allow sea water from outside to make its way back through the exhaust system and into the engine. To stop this, many boats have an extra loop in the exhaust system, known as a *swan-neck*. To guard against the possibility of waves pushing water up the exhaust pipe, some boats have a one-way flap covering the end of the pipe where it emerges from the hull; on some sailing yachts you may even find a hand-operated *gate valve* that seals the exhaust pipe completely when the engine is not being used.

The vital thing about any exhaust system is that it must not restrict the flow of exhaust gases beyond a certain limit, because if the exhaust can't get out of the cylinders, there will be no room for fresh air to get in. The effect is exactly the same as if the air filter were clogged: starved of oxygen, the engine will not be able to burn its fuel, so it will lose power and produce black smoke.

MORE POWER

Any engine is simply a device for converting the energy released from burning fuel into mechanical power. None of them are very good at it: well over 60 per cent of the energy released from the fuel is expended as heat and vibration, rather than as useful mechanical work. Engine designers are continually working to improve efficiency, but the fact remains that the power an engine can produce will always be limited by the rate at which it can burn fuel.

TURBOCHARGER

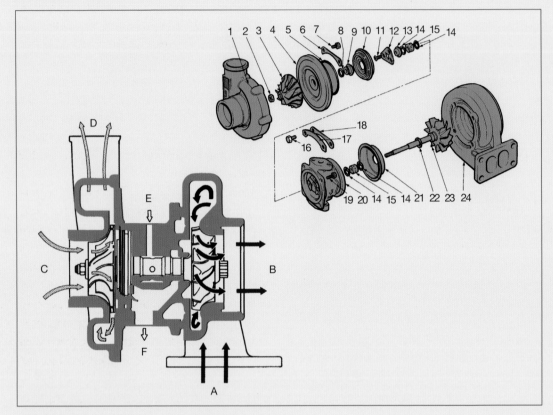

Fig 11 Working principle and parts of the turbocharger.

KEY

LOWER DRAWING

A. Hot exhaust gases from the manifold drive the turbine.

B. The exhaust gases then pass through the exhaust pipe/silencer to be cooled by a raw-water injection bend fitted after the outlet.

C. Air from the air cleaner is fed into the compressor.

D. The compressed air is fed through the air inlet manifold to the cylinders where it can burn an increased amount of fuel compared to a normally aspirated engine.

E. Lubricating oil inlet from engine's pressurised oil supply.

F. Lubricating oil returns to engine sump.

TOP DRAWING

1. Compressor-housing.
2. Lock-nut.
3. Compressor impeller.
4. Compressor diffuser.
5. 'SQ' ring seal.
6. Clamping plate.
7. Hex head screw and washer.
8. Seal (split ring).
9. Oil slinger.
10. Oil baffle.
11. Flat head cap screw.
12. Thrust bearing.
13. Thrust collar.
14. Circlip.
15. Bearing.
16. Hex head screw.
17. Lockplate.
18. Clamping plate.
19. Hex head set screw and washer.
20. Bearing housing.
21. Heat shield.
22. Seal (split ring).
23. Shaft and turbine wheel.
24. Turbine housing.

At the present state of development, a good rule of thumb is that every gallon of diesel fuel will produce about 20 hp for one hour – or 10 hp for two hours, or 100 hp for 12 minutes, and so on. So if you want an engine to develop 40 hp, for instance, it needs to burn about 2 gallons per hour.

It's relatively easy to squirt more fuel into the cylinder, but that alone won't produce more power, because every gram of fuel needs about 25 g of air in order to burn. So to burn more fuel, you have to get more air into the engine.

This can be achieved in various ways:

◆ **Bigger cylinders** This has the advantage of simplicity and relatively low cost.

◆ **More cylinders** Involves more complex castings, more valves and a more complicated fuel system, but tends to be smoother-running, and more responsive.

◆ **Run the engine faster** This pumps more air through the engine, without increasing its size or weight. Almost all modern diesels run faster than their counterparts of 20 years ago.

◆ **Force air into the cylinders** Done by pressurising it, to squeeze the equivalent of about 1¼ litres of air into each 1 litre of cylinder capacity.

The latter option has become very much more popular over the past 20 years or so, and is usually achieved by means of a blower called a *turbocharger* driven by a turbine built into the engine's exhaust system.

Unfortunately, turbochargers have to operate at high temperatures and at speeds in the order of 100,000 rpm – which give conservative marine engineers the heebie-jeebies, and produce a high-pitched whine that some people find offensive. Nevertheless, turbochargers are usually very reliable, and coax about 25 per cent more power out of an engine very efficiently, by winning back some of the energy that would otherwise be wasted in flow of hot exhaust gas.

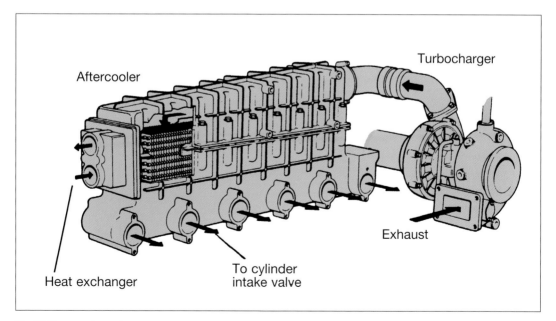

Fig 12 Charge air cooler.

One snag with a turbocharger is that pressurising air, especially by pumping it through a hot component like a turbocharger, raises its temperature; therefore it tries to expand – exactly the opposite of what the turbocharger is trying to achieve! To overcome this, many engines draw their air supply through a duct lined with pipes containing cool sea water called a *charge air cooler, intercooler* or *aftercooler*.

You can get some idea of how effective this is by looking at the specifications of an engine such as the 90 hp Mermaid Melody. With a turbocharger, the same engine becomes the 160 hp Turbo Melody; and with an intercooler as well, it's up to 200 hp – a 122 per cent increase in power for a 3 per cent increase in weight and 40 per cent increase in price.

VARIATIONS ON TURBOCHARGING

One application for which a turbocharged engine is not suitable is in a boat that spends most of its life operating at low speeds with only occasional, widely spaced bursts of high power. This is because at low power the exhaust flow won't be enough to operate the turbocharger. Exhaust gas flowing past the stationary turbocharger blades produces a build-up of soot, so when high power is called for, the clogged-up turbocharger can't work properly. As a result, the engine won't receive enough air to burn its fuel properly, so it will produce more oily soot that makes matters even worse.

There are various ways in which designers have brought the benefits of turbocharging to engines that have to operate at a wide range of speeds.

One method is to fit a smaller turbocharger, capable of operating even with the reduced flow of exhaust gas produced at low engine speeds. This, however, means that at high revs the turbocharger will be faced with more exhaust than it can cope with, so some of the exhaust has to be diverted away through a bypass arrangement called a *waste-gate*.

An alternative is to use a mechanically driven compressor called a *supercharger* at medium revs, allowing the turbocharger to take over as the engine speed increases.

... THINGS TO DO

1. BLACK SMOKE

Black smoke is almost certainly a symptom of problems in the air system.

a. Check to make sure that the engine room ventilation is adequate – that the engine compartment ventilation louvres aren't blocked by leaves or by the contents of lockers, for instance – and that the engine's air filter isn't clogged.

b. On turbocharged engines, look for loose hoses or leaks between the turbocharger and the engine itself.

c. Make sure the exhaust hose isn't blocked, squashed or damaged: bear in mind that flexible exhaust pipes can deteriorate in time, allowing their inner layers to collapse while the outside looks perfectly sound.

2. AIR FILTER

a. Clean or replace the air filter at least once a season. Unclip or unscrew the cover, and lift out the filter element. Paper elements should be replaced if they are dirty or damaged.

b. Wire gauze filters should be washed in paraffin or a solution of washing-up liquid in water, and allowed to dry. Inspect the filter for rust or loose strands, and replace it if necessary. Otherwise, dip it in clean engine oil and drain off the excess.

5 COOLING SYSTEM

The previous chapter mentioned that over 60 per cent of the energy produced by burning fuel in a diesel engine is wasted in the form of heat. That's almost inevitable: heat is needed to ignite each charge of fuel and air in the first place, and it's heat that expands the contents of the cylinder to drive the piston downwards. The piston sliding up and down inside the cylinder produces yet more heat by friction, as does the movement of the con rod on the crankshaft and the rotation of the crankshaft in the main bearings – anywhere, in fact, where metal moves against metal.

If all this heat were retained by the engine, it would get hotter and hotter, until it either set fire to the boat or welded some of its own parts together to become a useless lump of dead metal.

Very small engines have a large surface area compared to their volume and the heat they produce, so a lot of heat can be lost to the atmosphere by radiation – so lawnmowers, small motorbikes and light aircraft need no cooling system as such, other than fins to increase their surface area. It is very different for boats: their engines are normally larger, and are invariably tucked away in snug engine compartments. They are, however, blessed with a plentiful supply of water.

THE BASIC SYSTEM – RAW-WATER COOLING

Some of the simplest water cooling systems are found in small outboards such as the old British Seagull. Its cylinder is cylindrical, but it's inside a cube-shaped cylinder block, which leaves large open spaces between the walls of the cylinder and the outer walls of the block that are filled by sea water pumped up from the bottom of the drive leg. The water absorbs heat from the cylinder, and then escapes back to the sea through a hole in the casting, pushed out by more water coming up from the pump.

Components such as the piston and crankshaft don't have the advantage of being in direct contact with the cool sea water, so they get much hotter, but are kept down to a reasonable working temperature by being able to conduct heat away to the relatively cool block.

This kind of system is called *direct cooling*, or *raw-water cooling*, and is so simple, cheap and effective that it would be surprising if it wasn't also used in small diesels.

The main difference between a diesel's direct cooling system and that of an outboard is that the diesel's cooling water has to be pumped into the boat and back out again.

The way in is through a hole in the boat and a flexible hose. The hole has to be below the waterline, so any leaks from any part of the cooling system are potentially capable of sinking the boat. This makes a *seacock* essential, so as to be able to isolate the entire system from the sea.

If the system gets blocked accidentally, by weed or rubbish, the consequences are less dramatic, but are still potentially serious. To guard against this, the system should have a *raw-water filter*.

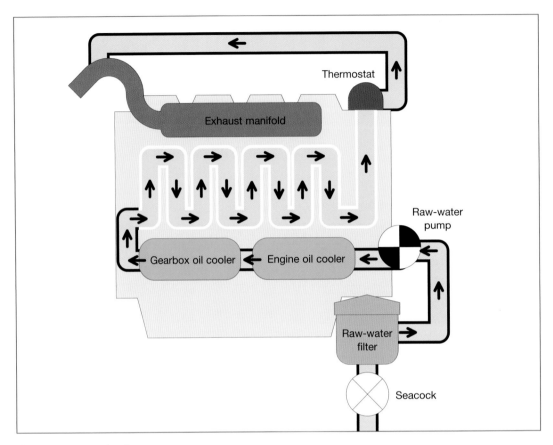

Fig 13 Raw-water circuit.

Once the water has done its job of cooling the engine, it can be discharged overboard through a hole in the topsides. Nowadays, though, it's much more common for it to be mixed with the engine's exhaust gas in the *injection bend*, where it cools and quietens the exhaust system. To reduce the risk of water from the cooling system flooding the exhaust manifold when the engine is not running, an anti-siphon valve is usually built in just before the injection bend.

THE THERMOSTAT

One drawback of raw-water cooling is that it can be too effective, especially when the engine is being started, or when it is running at low load. The engine needs some heat to ignite its fuel, so removing heat through the cooling system can be counterproductive.

To overcome this, and allow the engine to start and run at its most efficient temperature, most diesels are fitted with an automatic valve called a *thermostat*, which regulates the flow of cooling water.

The thermostat is usually mounted under a dome-shaped cover where the cooling water leaves the cylinder head. It's a simple component, whose only moving part is a circular trapdoor of thin metal, held shut by a spring. Under the trapdoor is a sealed capsule of wax or alcohol that expands as the temperature of the surrounding water rises until it overcomes the resistance of the spring and pushes the trapdoor open.

If the thermostat were 100 per cent effective at shutting off the water flow, there would be quite a build-up of pressure between the pump and the thermostat, so the thermostat has a small bypass hole to allow some water to flow when the thermostat is shut. Even if there's a separate bypass hose, the hole has an important role. Without it, an air lock could keep the cooling water away from the thermostat – thereby stopping it from opening until the temperature of the engine had already risen dangerously high.

It's worth bearing in mind that there is bound to be a slight difference between the temperature at which the thermostat opens and the temperature at which it closes, so if you watch the temperature gauge closely you may well see a slow and fairly regular rise and fall in engine temperature. This is nothing to worry about: just get used to the normal range of operating temperatures for your engine.

Thermostats can occasionally jam open or closed. If yours jams open, the immediate effect will be that the fluctuation of engine temperature stops, and the engine runs cooler than usual, burning more fuel but producing less power and more smoke.

A more serious problem arises if the thermostat jams shut: the bypass flow alone won't be enough to cool the engine, so it will overheat.

It's easy to test a thermostat by taking it out and putting it in a saucepan of water on the stove. As the water temperature rises, you should see the thermostat open. This should happen when the water is too hot for you to bear to put your hand in it, but well before it reaches boiling point.

If the thermostat has failed, a get-you-home solution is to break the wax capsule and spring away to allow the trapdoor to stay open.

Removing a thermostat.

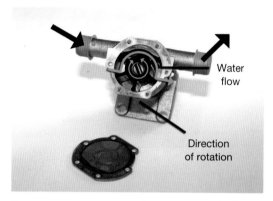

Water flow

Direction of rotation

Raw-water pump: note that the vanes rotate clockwise, and that the water flow is in the same direction (ie the 'long way round').

RAW-WATER PUMP

There are many different ways of pumping water for raw-water systems, but the most common by far is the 'flexible impeller' type of pump – often known by the trade name *Jabsco*.

The flexible impeller looks like a paddle wheel, with several flat blades or vanes sticking out from a central hub. It's a tight fit inside a cylindrical casing, and is made even tighter by a bulge in the wall of the casing, between the inlet and outlet pipes.

As the impeller turns, each vane in turn has to bend to get past the bulge. This reduces the space between the bent vane and the one in front. As the vane clears the bulge, it straightens out again, increasing the space between the two vanes

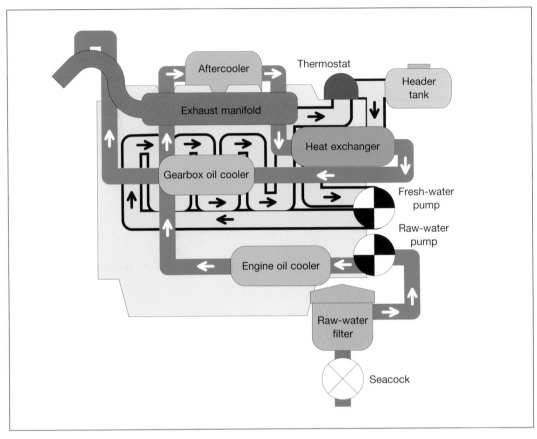

Fig 14 Fresh-water circuit.

and pulling water in from the inlet pipe. As the impeller continues to rotate, the water trapped between the two vanes is carried around with it, until it reaches the outlet pipe. At this point, the leading vane encounters the bulge in the casing and has to bend again to get past it. This reduces the space between the two vanes, and forces the trapped water into the outlet pipe.

ANODES

Warm sea water is ferociously corrosive, so an engine with raw-water cooling needs something to reduce the effect. Just as most boats have sacrificial zinc anodes below the waterline to protect

exposed metal parts, so do most raw-water-cooled engines. Engine anodes come in many shapes and sizes, though they are often in the form of rods, about the shape and size of a man's finger, which screw into holes in the engine block.

Sacrificial anodes are very effective, but are inconspicuous and easily forgotten, so do check the engine instruction manual to find out where they are and when they should be replaced.

INDIRECT COOLING

An alternative solution to the problem of corrosion is to keep sea water away from the engine altogether, and use fresh water – usually mixed with antifreeze as further protection against corrosion.

This is exactly the same as the way car, truck and tractor engines are cooled, so it is particularly common in engines over about 50 hp (which are almost invariably based on designs intended for use in vehicles).

Fresh-water cooling has other advantages besides reducing the risk of corrosion: it offers closer control of the engine's operating temperature, and allows it to run slightly hotter without salt deposits building up in the pipe-work. Both of these make the engine more efficient, so fresh-water cooling is gradually becoming more common even on engines as small as 10 hp.

The big difference between a boat engine and its stablemate in a car or truck is that a boat engine doesn't use an air-cooled radiator to cool the water that has cooled the engine. Instead, it uses a *heat exchanger*, made up of a bundle of small-bore tubes or thin hollow plates inside an outer casing. The fresh water flows through the casing, while raw (sea) water flows through the tubes or plates.

Fresh-water cooling, then, involves two sub-systems: a fresh-water system that cools the engine, and a raw-water system that cools the fresh water. For this reason, it's often known as *indirect cooling*.

Apart from the heat exchanger, the components involved in an indirect system are much the same as those that make up a raw-water system, because the raw-water side still has to have a seacock, filter, pump and injection bend. Only the thermostat is missing, because it is now part of the fresh-water system.

Two extra components are involved in the fresh-water side: a *header tank* to provide a reserve of cooling water and give room for the water to expand and contract as its temperature changes; and a *circulating pump* to drive water through the system.

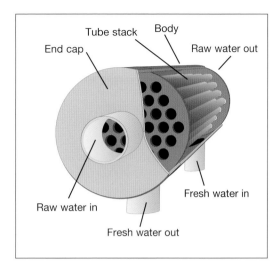

Fig 15 Heat exchanger.

The header tank is often combined with the heat exchanger, to form a substantial box-like component mounted high on the front of the engine. It's topped by something very much like a car's radiator cap and serving almost exactly the same purpose – keeping coolant inside the system even when it tries to escape as steam, but acting as a safety valve if the pressure rises too high.

Like a radiator cap, the header tank cap can eventually fail, when the sealing ring is damaged or when the spring loses its resilience. Either of these will lead to a steady loss of water, which could eventually lead to the engine overheating.

Doom and gloom merchants will tell you (quite correctly) that loss of water and overheating are among the symptoms of a blown cylinder head gasket. It could save you a lot of money if you try replacing the header tank cap before leaping to the assumption that the pessimists are right!

CIRCULATING PUMP

Compared with the raw-water pump, which may have to lift water from the sea, the fresh-water pump has the relatively simple task of creating a flow of water through an enclosed system. This means the pump itself can be the simpler *centrifugal* type, which is less prone to wear and tear.

The outer casing is dome-shaped, with the inlet pipe at its centre and the outlet pipe emerging from the edge. Inside, the impeller is virtually flat, but has curved vanes sticking up from its surface like the fan of a hover mower. As the impeller spins, the vanes set up a swirling movement of the water inside the casing. Centrifugal force, helped by the curvature of the blades, drives the water out into the outlet pipe, while more water rushes in through the inlet pipe to fill the space that would otherwise be left in the centre.

There is little to go wrong with a centrifugal pump until – after several thousand hours' running – the bearings that support its shaft start to wear, producing a high-pitched and almost continuous squeak. When this happens, it's a fairly simple job to rebuild the pump with new components and even easier to replace the whole thing.

SKIN COOLING

A variation on indirect cooling, popular in steel canal boats and some small commercial vessels, is known as skin cooling or by the somewhat misleading name of 'keel cooling'.

Essentially it replaces the heat exchanger with tubes or with a tank that is in direct contact with the side or bottom of the vessel. Coolant passing through the tank or tubes discharges its heat through the metal skin of the vessel, into the surrounding water.

Skin cooling systems require very little in the way of maintenance, but it is important to check the coolant level regularly, and to make sure any

Fresh-water circulating pump.

hoses are in good condition. Every couple of seasons or so, replace the coolant with a fresh mixture of antifreeze and water.

OIL COOLING

One tends to assume that an engine needs oil for much the same reason as a bicycle or a door hinge – to reduce friction so that things move or slide over each other more easily. That, however, is only one of its jobs: it also plays an important part in carrying heat away from the engine's most intimate working parts. Some engines, for instance, even have a system that squirts a jet of oil against the underside of each piston.

In a car, the heat that the oil has collected as it travels around the engine is dissipated from the sump, hanging down below the engine in the rush of air passing under the vehicle.

For obvious reasons this doesn't apply to marine engines, so many – particularly those over about 50 hp – have an *oil cooler*.

An oil cooler is another heat exchanger, similar to the main heat exchanger but smaller, that uses the engine's raw-water system in order to cool the oil. A second oil cooler is often used to cool the gearbox oil.

... THINGS TO DO

SAFETY FIRST

Remember that when the engine is warm, the fresh-water system may be full of very hot water or steam, and under pressure.

The raw-water system is directly connected to the sea. Any leak is potentially capable of sinking the boat.

1. CLEARING THE RAW-WATER FILTER

The raw-water filter should be checked, and cleared if necessary, each day that the engine is to be used, and whenever there is an unusual rise in engine temperature.

If your filter has a transparent cover, putting a table-tennis ball inside can save time and trouble because movement of the table-tennis ball is a clear sign that water is flowing through the filter. Otherwise, when you start the engine, get into the habit of checking that water is coming out of the exhaust pipe.

Raw-water filters differ in design and construction. In general, though, the procedure is:

a. Shut the raw-water seacock.

b. Remove the cover: this may involve undoing several nuts, unscrewing the cap as though you were opening a jam jar, or releasing a clamp.

c. Remove the filter element – usually a cylinder of perforated sheet metal, wire gauze, or a net of nylon mesh covering a metal frame – and clear out any weed or debris.

d. Put the filter back, making sure that any locating studs fit into their notches, and that the top of the filter is at the same level as it was before.

e. Replace the filter cover, making sure that it is screwed down hand-tight.

f. Open the seacock and inspect for leaks around the cover. **Don't be tempted to leave this while you do your other daily checks – it's too easy to start the engine with the seacock closed!**

2. CHECKING THE HEADER TANK

If your engine has an indirect cooling system, the level of water in the header tank should also be checked whenever you check the raw-water filter.

a. Unscrew the header tank cap. If the engine is warm, protect yourself by covering it with several layers of cloth (such as an old towel), and unscrew it very slowly to allow any pressure to be released gradually. Some types have a 'bayonet' fitting: these have to be pressed down against the spring pressure before they can be unscrewed, but take only a quarter turn to release: some have a two-stage unscrewing action that allows them to be partly unscrewed to release any pressure, then require a second push-and-twist action to release them completely.

b. Most manufacturers recommend that the water level should be between 1 and 3 in (25 and 76 mm) below the top of the tank: in general, if you can

1b

2b

touch the water with your finger, it's full enough. If not, top it up with clean fresh water mixed with antifreeze. Replace the cap.

3. REPLACING THE RAW-WATER PUMP IMPELLER

Although 'Jabsco'-type pumps are virtually standard, there are many different models, so it's a good idea to carry at least one spare on board, because the impeller will very quickly be damaged if the pump is run dry. If the flow of cooling water stops, or the engine shows signs of overheating, check the filter first, then the pump impeller.

a. Undo the screws holding the pump's front cover in place, and remove it. Peel away the remains of the paper gasket that may be stuck to the cover, or the body, or a bit of both.

b. Pull out the impeller with a pair of pliers. If it won't come out, or if you have no suitable pliers available, it can be prised out using two screwdrivers, but be very careful not to damage the softer metal of the pump body.

c. If the impeller has disintegrated, try to piece it together so as to be certain there are no missing

pieces wandering around the cooling system, where they could cause blockages later. If there are any missing pieces, try to find them if you can: dismantle the pipe-work between the pump and the heat exchanger to see if they are stuck at a bend, or look in the heat exchanger itself.

d. Smear the new impeller with washing-up liquid, and slide it on to its shaft, making sure that the vanes are bent the right way and that it is properly located on the drive key or pin that protrudes from the shaft. Notice (see photo on page 29) that the water always takes the 'long way round' in its trip from the inlet pipe to the outlet and that the vanes trail backwards like the sparks from a Catherine wheel.

e. Use a little water or washing-up liquid to temporarily stick the new gasket that is supplied with the impeller in place on the pump body, and then replace the cover plate.

4. REPLACING INTERNAL ANODES

Check with your engine instruction manual to see how and when to replace internal anodes.

3a

3b

3c

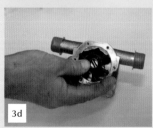

3d

3e

4

6

OIL SYSTEM

One of the quickest and most sure-fire ways to wreck an engine is to run it without oil, because even a smooth or polished surface has minute imperfections called asperites. These jagged spikes and ridges may only be a millionth of an inch high, but as metal moves against metal, the asperites on one surface collide and interlock with those on the other, and then have to bend or break in order to allow the movement to continue.

The cumulative effect of repeatedly bending or breaking thousands of asperites soon produces visible damage to the surface, known as wear. The effort expended in doing so is friction.

Each collision also generates heat, with local temperatures sometimes rising to as much as 1,600° C – enough, in severe cases, to weld the surfaces together.

Oil solves the problem by separating the two metal surfaces with a thin layer of fluid, and filling the tiny valleys between the asperites: reducing wear, allowing moving parts to move more freely, and stopping them from welding themselves together.

CLEANS, COOLS AND PROTECTS

Lubrication is undoubtedly the oil's main job, but it has a number of subsidiary functions. These were neatly summed up by one of the oil companies whose advertising slogan, at one time, was 'cleans, cools and protects'.

♦ It cleans the engine by flushing away tiny particles of carbon or metal, and neutralising the acids produced from burning fuel.

♦ It cools the engine by carrying heat away from hot spots that can't be reached by the engine's water cooling system, such as the pistons and main bearings.

♦ It protects the engine by covering the metal parts to exclude air and moisture that would otherwise cause corrosion.

One significant job that the advertising agency forgot is that oil also helps to create a gas-tight seal between – for instance – the piston rings and the cylinder walls.

PRESSURISED OIL SYSTEMS

A few very simple engines have no oil system at all. Two-stroke petrol engines, for instance, run on a mixture of oil and petrol, and rely on the engine's demand for air to pull the petrol-oil mixture through the areas where oil is needed. Some small four-stroke engines rely exclusively on 'splash-feed' lubrication, in which a spike or paddle, protruding from the bottom of the con rod, flings oil around the inside of the crankcase, the bottom of the cylinder, and the underside of the piston.

You don't have to move far up the scale, though, to reach engines in which pressurised oil systems are standard. They use an oil pump to lift oil out of the *sump* and through a filter, before pushing it through a maze of *oilways* to the crankshaft, camshaft and rocker bearings; to the con rods and pistons; and out to ancillaries such as the turbocharger and fuel pump. Gravity then returns the 'used' oil to the sump.

Most of the lubrication system – like the pistons and main bearings that it serves – are deep inside the engine, and out of reach of a limited onboard tool kit. User maintenance is confined to making sure that the engine has a good supply of clean oil by topping up and changing the oil at regular intervals, and changing the filter.

OIL SYSTEM

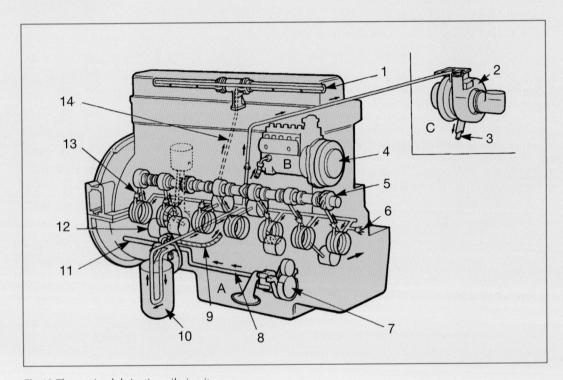

Fig 16 The engine lubricating oil circuit.

KEY

A. The sump and coarse filter.

1. Rocker shaft.

2. Turbocharger (some petrol and diesel engines only (C)).

3. Oil drain to oil pan from turbocharger.

4. Lubrication for fuel injection pump (some 'in-line' diesel pumps only (B)).

5. Camshaft bearings.

6. Spray to timing gear.

7. Oil pump.

8. Feed pipe to filter and thence to main oil gallery.

9. Feed to diesel fuel in-line injection pump (see 4 above).

10. Oil filter.

11. Main oil gallery.

12. Big end bearings.

13. Main bearings and piston cooling nozzles.

14. Metered oil feed to rockers and rocker shaft.

In the longer term, it is a good idea to keep an eye on the oil pressure gauge. Some internal wear is inevitable, so as time goes by, the gaps between some of the moving parts will increase, making it easier for oil to seep away. This doesn't just mean that the lubrication around the affected part will be less effective: it also means that less oil will reach other components, leaving you with an escalating trail of damage throughout the entire engine.

Good operating practices delay the onset of wear: in particular it pays to remember that until the engine has been running for a few moments, it will be dependent on the oil that has been left clinging to its bearing surfaces. If the engine has been idle for very long, those surfaces may be almost completely dry, so if it is started with full 'throttle' and roars into life as soon as the key is turned, a lot of wear will take place before any new oil has had a chance to reach the parts it is supposed to protect.

OIL GRADES AND CLASSES

It is pretty obvious that if the oil is to do its job of separating two moving parts, there has to be a gap between them that the oil can fill. This, however, means that if the oil were perfectly fluid it would simply escape through the gap, so to be effective an oil needs a certain *viscosity*, or 'thickness'.

This seems to imply that a 'thick', viscous oil is better than a 'thin' one, but that is certainly not the case: viscosity is an indication of the friction between the molecules of the oil itself, so a very viscous oil makes starting difficult, wastes power and generates extra heat.

In other words, you need to choose an oil of the right viscosity for your engine.

There are lots of different ways of measuring viscosity but, to make life relatively simple,

oils are now graded according to a system of numbers devised by the American Society of Automotive Engineers, in which the higher the number, the thicker the oil. Your engine manual may specify, for instance, that it needs an oil grade 'SAE 40'.

The picture is made slightly more complicated by the fact that oils become less viscous as they warm up, so an oil that is right at normal operating temperature may be very much too thick for easy starting in the depths of winter. To overcome this, it was once common practice to use a much 'lighter' oil in winter, and to accept increased wear as a penalty that had to be paid. SAE catered for this by introducing a second series of 'Winter' grades, such as SAE 10W.

Oil technology has advanced enormously since the SAE grades were introduced. Now, additives mixed with the oil make it much less susceptible to changes in temperature. This means that most modern engine oils can be used in summer and winter alike, and therefore have summer and winter SAE grades shown together, such as SAE 20W/50 or SAE 15W/40.

Even 'ordinary' engine oils nowadays contain a cocktail of other additives intended to enhance particular aspects of their performance. Inevitably this means that some oils are 'better' or 'worse' than others, so various bodies have introduced performance standards to identify oils that are suitable for particular jobs.

The most widespread of these classification systems was developed by the American Petroleum Institute (API), which assesses an oil's performance in each of two categories: S, for spark ignition (petrol) engines; and C, for compression ignition (diesel) engines. As time has gone by, the capabilities of the oil producers and the demands of the engine manufacturers have increased, so now there are a range of API classifications from SA/CA (the oldest and obsolete) up to SN and CK.

Most oils now meet SJ/CH or SK/CI specifications, and are perfectly suitable for use in most engines, but if you are faced with an unfamiliar brand it's as well to check the quality designators printed on the can and to check, if necessary, that it is specified as suitable for a turbocharged engine.

... THINGS TO DO

SAFETY FIRST

The additives that make modern oils better for your engine make them worse for you. Take care to avoid unnecessary or prolonged contact with engine oil – new or used.

1. OIL LEVEL

Check the oil level each day that the engine is to be used. It is quite normal for an engine to 'use' a certain amount of oil.

a. Withdraw the dipstick, wipe it with a dry rag and then put it back, making sure it is pushed fully home. Pull it out again, and look at the oil level, which should be between the 'max' and 'min' marks. Then replace the dipstick.

b. If necessary, top up the oil by pouring oil in through the filler cap – usually located on top of the rocker cover. Leave the engine for a few seconds for the new oil to drain down before re-checking the level.

2. CHANGING THE OIL FILTER

The oil and filter should be changed at the end of each season, or after about 200 hours' use.

First run the engine up to operating temperature, then protect the area around the filter from spillages. Try to avoid contact with the used oil.

a. 'Spin-on' filters are best removed by unscrewing with a strap or chain wrench.

b. If that isn't available or doesn't work, drive a large screwdriver through the canister, just off-centre, and use it as a lever.

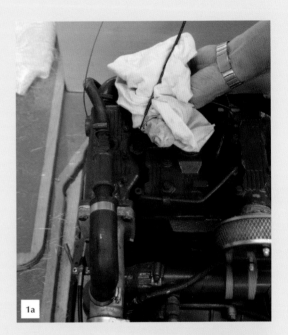

1a

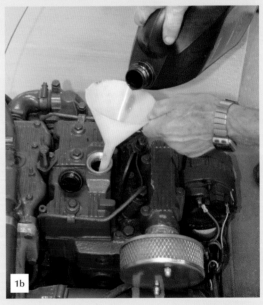

1b

2a

2c

f. Make sure the old rubber sealing ring isn't stuck to the filter head, and replace it with the new one supplied with the filter.

g. Replace the complete assembly, making sure the filter canister is correctly seated, and tighten the retaining bolt.

h. Change the oil, then run the engine at tick-over for a few minutes to inspect for leaks around the filter. Even if no leaks appear, check the oil level and top it up, because some oil will be retained in the filter.

3. CHANGING ENGINE OIL

There's no point changing the filter without changing the oil, and if your engine has a sump pump, this is a simple matter of pumping the oil into a suitable container. If not, you will have to resort to other methods:

a. You will have to insert a small tube down the dipstick hole, connected to a pump, to suck the oil out.

b. Top up with the right grade and quantity of oil.

3a

c. Smear the sealing ring with a thin film of fresh oil. Spin the filter on until the sealing ring just touches the filter head, then tighten it another half turn by hand. Do not over-tighten it.

d. Cartridge-type filters are usually secured by a single bolt.

e. Remove the cartridge, and replace it, making sure that the various springs and washers are replaced in the correct order, and that the filter is the right way up.

ELECTRICAL SYSTEM

7

One of the great virtues of a diesel engine is that it doesn't need electricity to start or run. Small diesels can be started by hand, bigger ones by coiled springs, and the biggest of all by compressed air or separate starter engines.

An electric motor, though, is such a convenient way of doing the job that almost all marine diesels now have an electrical system of some description.

THE BASIC SYSTEM

Diesels intended for hand starting usually have *decompression levers*, which make it a lot easier to build up momentum in the heavy flywheel by opening the valves in the cylinder head to release the pressure in the cylinder.

Even so, hand starting a diesel can still be hard work, so it shouldn't come as any surprise to find that it takes a lot of electrical power to do the same thing – especially as electric-start engines don't usually have decompression levers. Compared with any other electrical equipment on board, the amount of power involved is huge: the initial surge of current may be in the order of 1,000 amps. It quickly drops away to about 200 amps as the motor picks up speed, but it's still asking a lot to expect any battery to start an engine more than once or twice without needing to be recharged.

This means that as well as the *starter motor* itself and a switch to control it, even the most basic electrical system is likely to include a dynamo or *alternator* to recharge the battery, and a *regulator* to stop the battery being overcharged. For indirect injection engines, there is usually also some kind of *cold starting aid* to make the starter motor's job a little easier by warming the air in the cylinders.

No ordinary switch can cope with the sort of currents involved in the starter circuit, so the 'switch' that controls the starter motor is actually a relay or *solenoid* – a remote-controlled switch operated by another switch on the control panel. The control panel also needs a warning light or buzzer to attract attention to any failure of the electrical system, so the complete system ends up looking something like that in Fig 17.

MAKING ELECTRICITY

Generators and motors both depend on the close link between electricity and magnetism, which can be summed up by saying that if you move a wire in a magnetic field you'll create electricity, and if you pass electricity through a wire you'll create magnetism.

Fig 18A shows how this might be applied to make electricity: a coil of wire, wound around a central core, is spinning in the magnetic field between two magnets. As it rotates, the wires that make up the coil move through the magnetic field, to generate an electric current.

Fig 18B shows one way of getting the electricity out of the coil: the ends of the wire are connected

BASIC ELECTRICAL CIRCUIT

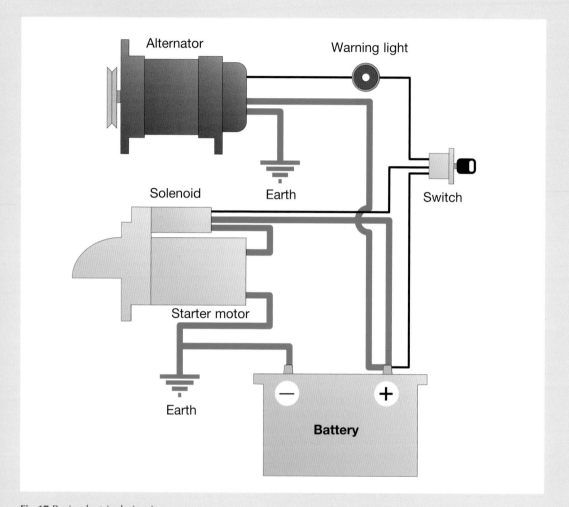

Fig 17 Basic electrical circuit.

Tracing an engine's wiring loom can be complicated because some wires do two or more jobs, and it is not possible to see exactly what is going on inside some components.

Here, the battery supplies power to the starter solenoid, which – when activated – passes it on to the starter. The starter returns the current to the battery through the engine block itself, shown by the 'earth' symbol.

The solenoid is activated by the key switch, which draws power from the battery through a thinner wire and feeds it to the solenoid. The solenoid current returns to the battery through the starter's earth connection.

The alternator creates current that flows straight to the battery and is returned through the engine block. A thinner wire links the alternator terminal labelled 'IND' to the warning light, which is connected to the battery through the key switch. So long as the alternator is charging, no current flows through this wire, but if the alternator fails, current flows from the battery, through the switch and warning light, and returns to the earth.

to brass *slip rings*, fixed to the spinning shaft. Self-lubricating carbon *brushes* press against the slip rings to collect the electricity, and wires take it away to the rest of the system.

This is simple, but for practical purposes it has a major problem: because the coil is passing up through the magnetic field on one side, then down through it on the other, the current created in it is regularly changing direction.

Batteries can't handle *alternating current* (AC): they need a one-way current (*direct current*, or DC). One possible solution is shown in Fig 18C. Instead of a pair of slip rings there is only one, split in half along its axis. Every time the current flow reverses, the connection to the outside world reverses as well, producing an intermittent but one-way current.

DYNAMOS

A real dynamo looks much more complicated than the contraption shown in Fig 18, but works in exactly the same way.

Instead of a single coil of wire, it has several, each wound around its own soft iron core. This, in turn, means that the two-piece 'slip ring' of the rudimentary dynamo has to be replaced by a *commutator* made up of several segments – one pair for each coil. This produces a smoother flow of current, because as the flow generated by one coil is reducing, the flow from another is increasing.

The other major refinement in a real dynamo is that instead of using permanent magnets to create the magnetic field, it uses electromagnets. They consist of soft iron cores bolted to the casing of the dynamo, and wrapped in coils of copper wire called the *field windings*. Passing an electric current through the field windings creates the magnetism.

The virtue of using electricity to make more electricity is that it allows the intensity

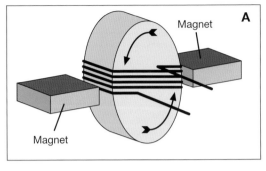

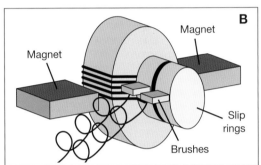

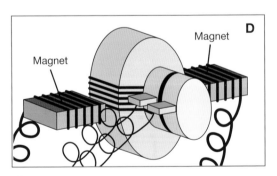

Fig 18

A. A simple generator.
B. An AC generator.
C. A simple dynamo.
D. Electromagnets allow control of output.

of the magnetic field to be varied. A separate component called the *regulator* monitors the output of the dynamo and interrupts the power supplied to the field windings if the output rises too high, allowing the dynamo's output to be controlled without having to adjust its speed.

ALTERNATORS

Dynamos are something of a rarity nowadays, having been almost completely replaced by a more efficient kind of generator called an *alternator*.

Mechanically, an alternator can be regarded as a dynamo turned inside out, with its field windings mounted on a spinning *rotor* in the middle, surrounded by a fixed ring called the *stator*, which carries the generating coils.

This arrangement leaves the coils stationary, but subjected to a spinning magnetic field. Having stationary generating coils means that an alternator can't use a commutator arrangement to convert AC into DC: instead, it uses a solid-state electronic *rectifier* and regulator, usually built in under a plastic cover at one end of the alternator body.

As well as being more efficient than dynamos, alternators are generally more reliable and require little or no maintenance. Their main weakness is that their electronic components burn out very quickly if they are kept running with nowhere for the electricity to go, so it is essential to make sure

Power is generated by the alternator.

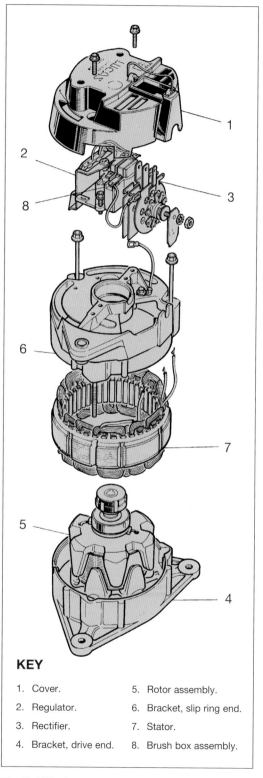

KEY

1. Cover.	5. Rotor assembly.
2. Regulator.	6. Bracket, slip ring end.
3. Rectifier.	7. Stator.
4. Bracket, drive end.	8. Brush box assembly.

Fig 19 ACR alternator.

never to disconnect an alternator while the engine is running – by switching the battery isolator switch off, for instance. If your installation allows you to stop the engine by turning a key, make sure it really has stopped completely before switching the key to the 'off' position.

STARTER MOTORS

Structurally, an electric motor is very much like the rudimentary dynamo in Fig 18: it too has coils of wire mounted on a central shaft called the *armature*, and magnets bolted to the inside of the casing to create a magnetic field.

The big difference is that a motor uses electricity instead of creating it! When a current is passed through the armature coils, it creates a magnetic field that interacts with the magnetic field set up by the field windings in such a way that the armature is forced to rotate.

DYNASTARTS

A motor and a dynamo are so similar to each other that at one time there was a trend towards using a single component, called a *Dynastart*, to do both jobs. Unfortunately a good dynamo doesn't make a good motor or vice versa, so this rather neat idea has now gone out of fashion.

The starter motor.

BATTERIES

We can't create energy: we can only convert it from one form into another. That is what happens inside the engine itself: chemical energy is released from the burning fuel and converted into mechanical energy. The alternator then converts some of that mechanical energy into electrical energy. Electrical energy, however, can't be stored: to achieve that effect, the battery has to convert it back into chemical energy.

There are lots of ways of doing this, including the lightweight nickel-cadmium (NiCad) and nickel-metal hydride (NiMH) batteries used for mobile phones and hand-held radios. For the time being, however, these are too expensive to be used to store large amounts of power, so boats almost invariably use relatively cheap, low-tech *lead-acid* batteries.

The working part of a fully charged lead-acid battery is a stack of lead plates, interleaved with layers of lead peroxide and porous separators and surrounded by sulphuric acid. The acid tries to convert both sets of plates into lead sulphate while converting itself into water by rearranging the electrical charges that hold the molecules together. In the process, it creates an electric current between the positive peroxide plates and the negative lead plates.

Eventually, the sulphuric acid becomes so diluted that the reaction stops. The battery is then described as 'flat'. The beauty of the lead-acid battery, however, is that the whole process can be reversed by pumping electricity through it in the opposite direction. This converts the lead sulphate back into lead and lead peroxide, while the left-over sulphate turns the water back into sulphuric acid.

The system isn't absolutely perfect, though. For one thing, the conversion process is never 100 per cent completed: repeated charging and discharging leaves some unconverted lead

STARTER MOTOR

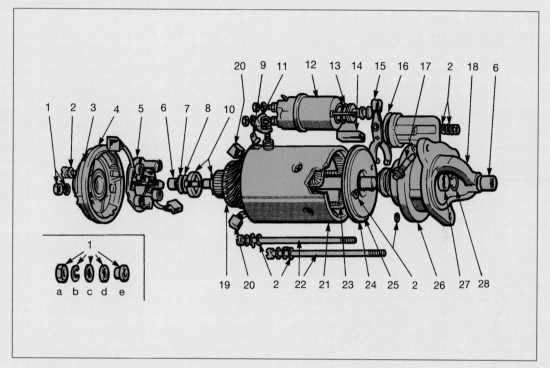

Fig 20 The Lucas M50 starter motor.

KEY

1. Nut and spring washer.

2. Sealing washers.

3. Commutator end cover.

4. Sealing ring.

5. Brush gear assembly comprising earth brushes and springs. Marine version has insulated negative brushes.

6. Bearing bush.

7. Fibre washer.

8. Steel thrust washer.

9. Flexible link.

10. Brake shoe and cross peg.

11. Copper link.

12. Solenoid unit.

13. Return spring.

14. Sealing grommet which is deleted on the marine version.

15. Engagement lever.

16. Gasket.

17. Eccentric pivot pin.

18. Drive and fixing bracket.

19. Armature.

20. Insulated brushes – field coils.

21. Yoke.

22. Through bolts.

23. Field coils.

24. Sealing ring.

25. Intermediate ring.

26. Drive assembly.

27. Thrust collar.

28. Jump ring.

Inset for marine version (a) Nut, (b) Plain washer, (c) Insulated washer for outside cover, (d) Insulated bush for inside cover.

sulphate on the plates, gradually reducing the battery's ability to 'hold its charge'. Eventually, the accumulated lead sulphate flakes away, to lie useless in the bottom of the casing.

Another problem is that passing electricity through water – or through a solution of sulphuric acid in water – causes a process known as electrolysis, which breaks down the water into hydrogen and oxygen. For safety reasons, this potentially explosive mixture of gases should be vented overboard, but it's important to appreciate that by doing so you're effectively discharging water from the battery, which will have to be topped up with distilled or de-ionised water from time to time.

The demands of starting an engine are very different from those imposed by 'ship's services' such as lighting and navigation equipment: the starter motor demands a lot of current for a few seconds at a time, while domestic and navigation equipment draws a relatively tiny current for hours at a stretch. Ideally, these contrasting requirements call for two different types of battery: a *heavy duty* battery for engine starting, and a *deep cycle* or *traction* battery for domestic loads. Structurally, the two types differ mainly in the number and thickness of their plates: a heavy duty battery has a large number of relatively thin plates, in order to expose as much surface area to the acid as possible, while a deep cycle battery has a smaller number of thicker plates in order to withstand the long-term effects of sulphation caused by being repeatedly discharged almost to the stage of being 'flat'. *'Marine'* batteries are a halfway house, intended mainly for small boats on which one battery may have to do both jobs.

FUSES AND CIRCUIT BREAKERS

Whenever an electric current passes through a wire it creates heat – and the heavier the current

or the thinner the wire, the hotter it is likely to get. Taken to extremes, this could melt the insulation around the wire, or even start a fire. To stop this happening, designers should specify cables that are big enough to allow heat to radiate away from the wire naturally.

There is still an element of risk, though, because the current could – for all sorts of reasons – increase to levels far greater than the designer envisaged. To make sure this doesn't happen, any properly installed electrical system incorporates some kind of excess current protection – either a *fuse* or a *circuit breaker.*

A fuse is a short piece of thin wire built into the system as a weak link that will heat up and break before the rest of the system suffers. The commonest type is similar to the fuse in a domestic plug, consisting of a short length of wire inside a glass tube. An increasingly common alternative is a small plastic peg with a strip of thin metal on the outside acting as the fuse. Both of these types have to be replaced as a complete unit when they 'blow'. The much rarer alternative consists of a piece of replaceable wire mounted in a porcelain carrier, similar to that in an old-fashioned domestic fuse box.

Circuit breakers are taking over from fuses because although they're more expensive, you don't have to carry spares. Essentially, a circuit breaker is an automatic switch, which uses either the solenoid principle or the heating effect to switch itself off if the current flowing through it becomes dangerously high.

Fuses and circuit breakers occasionally 'blow' for no apparent reason. Although this is a nuisance, it's essential not to give in to the temptation to replace a fuse with a bigger one, or to stop the breaker tripping with string or sticky tape! Find out why it's blowing, and cure the cause, rather than disabling your protection just when it's doing its job!

SOLENOIDS

A *solenoid* is an electrically operated switch consisting of a coil of wire around a metal plunger. When electricity flows through the coil it becomes a magnet, and pulls the metal plunger into itself. This pulling action has all sorts of purposes on an engine: it can, for instance, be used to shut off the fuel supply, allowing the engine to be stopped from the dashboard, or to operate the heavy duty switch that allows current to flow between the battery and starter motor.

On most small petrol engines, that is all the starter solenoid does – it just allows current to flow to the motor. The physical connection between the motor and the engine is achieved by an arrangement known as a *bendix*, made up of a cog (called a *pinion*) that can slide along a spiral groove machined into the motor's shaft. While the motor is stopped, a spring holds the pinion down towards the starter body by a spring. As the starter starts to turn, the pinion takes time to catch up so for a fraction of a second it is turning more slowly than the shaft on which it is mounted. As a result, the spiral groove in the shaft screws its way through the pinion, forcing it away from the starter body to engage with a circle of matching teeth (called the *ring gear*) on the engine's flywheel. As soon as the engine starts, the ring gear drives the pinion faster than the motor is turning, so the opposite happens: the pinion screws its way back down the starter shaft to disengage itself from the ring gear.

Diesels generally have a more rugged set-up called a *pre-engaged starter*. Here, a solenoid does two jobs. The first part of its travel is used to push the starter pinion into engagement with the ring gear: only when the pinion is engaged can the plunger complete its movement to allow current to flow to the motor.

... THINGS TO DO

SAFETY

The low voltages used in boats' electrical systems are safer than high-voltage mains electricity, but can still create heat or sparks that could potentially start a fire. Use the isolator switch to disconnect the battery before working on or near the starter or alternator, and before physically disconnecting the battery terminals.

Never allow metal objects such as tools, jewellery or watch straps to touch both terminals of a battery simultaneously. Ideally, keep the terminals covered when working near a battery, and keep one terminal covered while you are working on the other.

Batteries give off explosive gases while being charged, so the battery compartment should be well ventilated, with the vent high up (to cope with gases that are lighter than air). Don't smoke near a battery, and avoid creating sparks.

Battery acid is highly corrosive. Use goggles and gloves, especially when cleaning battery terminals, and avoid spills. If you get splashed use plenty of water to wash it off immediately.

1. CHECKING OR REPLACING DRIVE BELTS

The alternator or dynamo is usually driven by a *vee-belt* from the crankshaft pulley. The same belt may be used to drive the water pump, so if it is so loose that it slips, the engine will overheat. If it is too tight, it will put a sideways load on the alternator and water pump bearings, causing increased wear and making the belt itself more likely to break.

a. Once a season or after every 100 hours' running, inspect the belt for cuts, fraying or oil, and check the tension by pressing firmly on the middle of the longest section of the belt: it should deflect by approximately half an inch.

1a

b. Adjust the belt if necessary by loosening the bolts that hold the alternator on to the engine, without removing them altogether. Then slacken the bolt holding the supporting bracket to the engine, followed by the one holding the support on to the alternator. Use a short piece of wood such as the handle of a hammer to lever the alternator away from the engine until the belt is tight enough, then tighten all four bolts in the reverse order.

c. To replace a worn, greasy or broken belt, slacken all the alternator bolts, and swing the alternator so that it is as close to the engine as possible. Use a degreaser such as Gunk or Jizer to clean the pulleys and remove any rust by rubbing with wet-or-dry abrasive or wire wool.

d. Fit the new belt over the crankshaft pulley, water pump and idler wheel (if there is one), then up to the alternator. Guide it into place as best you can, then use a spanner on the crankshaft pulley to turn the engine while easing the belt into its groove in the alternator pulley. Adjust the belt tension, and check it again after a few hours' running.

1b

1d

2. LOOKING AFTER YOUR BATTERY

All electrical connections need to be clean and tight, but one of the commonest causes of starting problems is corrosion around the battery terminals.

a. Slacken the clamp that secures the cables to the battery, and twist it to remove it from the terminal post. Flush off any white or cream-coloured 'fur' – acidic crystals of electrolyte – with plenty of very hot water before cleaning the terminals and posts with a wire brush or emery cloth.

2a

b. Lightly smear the terminal posts with petroleum jelly, and wipe off any excess before replacing the terminal and tightening the clamp. Finally, smear the whole terminal with petroleum jelly to keep out moisture.

2b

c. Over time, electrolysis and evaporation remove water from the battery, which has to be replaced by pouring distilled water into each cell of the battery until it just covers the top edges of the plates. Some modern batteries have specially designed fillers intended to minimise the loss of electrolyte. Instructions for these are usually given on the battery casing.

d. If a battery is to be stored or left unused for more than a month or when there is any risk of freezing, it should be left disconnected and fully charged: a voltmeter connected between the + and – terminals should read no less than about 2.1 volts per cell, so a 6-cell, 12-volt battery should show at least 12.5 volts.

2c

GEARBOXES

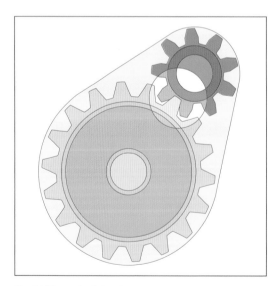

8

The overwhelming majority of engine-driven craft use propellers to convert the power of the engine to useful work. There are plenty of alternatives which may have advantages for specific applications, but propeller systems are good all-rounders that are reasonably cheap, simple, reliable, efficient and easy to use.

They suffer, however, from one potentially significant drawback, which is that a large slow-turning propeller is generally more efficient than a small, fast-spinning one. Even on small pleasure craft, where the maximum size of propeller is often limited by hull shape, the optimum shaft speed is usually in the order of 1,000 rpm – only a half or a quarter of the speed of the engine.

Running the engine more slowly isn't the answer, because you'd just lose most of the power you've paid for by buying it in the first place. What you need is something that will reduce the shaft speed but increase its *torque* (turning effort).

It's also very useful to be able to reverse the direction of rotation, to provide astern power to stop the boat or make it go backwards, or to fit counter-rotating propellers on a twin-screw boat. Again, there are alternatives, such as variable-pitch propellers whose blades swivel on the hub, scoop-like deflectors, or even engines that can be stopped and restarted in the opposite direction, but by far the most popular arrangement is a reversing gearbox.

BASIC PRINCIPLES

The diagram (Fig 21) shows two gearwheels, whose teeth mesh together so that as one turns, the other has to turn as well. The smaller gear has 9 teeth, so if it is turning at 1,000 rpm, its teeth are moving at 9,000 teeth per minute. The larger wheel is twice the size, and has 18 teeth, so although its teeth must also be moving at 9,000 teeth per minute, that means only 500 rpm. Notice, too, that if the smaller wheel is turning clockwise, the larger wheel must be turning anticlockwise.

Now imagine that you're using a spanner a foot long to turn the smaller wheel. If you apply an effort of 10 lb to the end of the spanner, you're applying a torque of 10 lb/ft (10 lb at a

Fig 21 The principle of gears.

radius of 1 ft). The gearwheel is much smaller – let's say it has an effective radius of 1 in. That means its teeth must be pushing on the teeth of the other wheel with a force equivalent to 120 lb (120 lb at a radius of $^1/_{12}$ ft = 120 x $^1/_{12}$ = 10 lb/ft). The larger wheel has an effective radius of 2 in, so a force of 120 lb to its teeth corresponds to a torque of 20 lb/ft (120 x $^2/_{12}$ = 20 lb/ft).

In other words, by using a 9-tooth wheel to drive an 18-tooth wheel, we've:

♦ halved the speed
♦ doubled the torque, and
♦ reversed the direction of rotation.

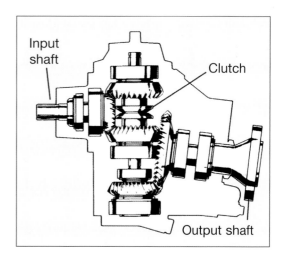

Fig 22 Volvo MS2 gear train.

A SIMPLE GEARBOX

Real gearboxes look more complicated, but depend on exactly this principle. In fact, the main difference between the simple gear train in Fig 21 and the Volvo MS2 shown in Fig 22 is that the MS2 uses cone-shaped bevel gears, so that although the input shaft is horizontal, the central driven shaft is vertical.

The motive for this arrangement is that it provides a simple way of achieving a choice of ahead or astern gear.

The bevel gear on the input shaft turns two slightly larger bevel gears that spin freely on the vertical shaft. One is driven by the top of the input gear and the other by the bottom of it, so they rotate in opposite directions, but they are the same size as each other so they turn at the same speed.

Between the two gears is a sliding clutch assembly, shaped like two shallow cones mounted base-to-base on the shaft. Ridges called *splines* on the shaft and matching grooves in the clutch ensure that it can easily slide up and down, but can't turn without turning the shaft as well. Moving the gear lever slides the clutch up

or down so that one of the cones engages into a matching hollow in one of the spinning gears, which locks that gear on to the shaft.

At the bottom of the shaft, a relatively small bevel gear engages with a much larger gear on the output shaft, to give a horizontal output, at a much lower speed of rotation.

LAYSHAFT GEARBOXES

Although the MS2 and its derivatives are common on Volvo engines up to about 200 hp, it's somewhat unusual in using a vertical shaft. A more widespread arrangement is the *layshaft gearbox*, like that shown in Fig 23.

In this case, a small gear on the input shaft turns a larger gear on the *layshaft*. Like the gears on the vertical shaft of the MS2, this is free to spin around the layshaft, but can be locked on to it by a clutch. If the input shaft is turning clockwise and the clutch is engaged, the layshaft turns anticlockwise and more slowly. At the other end of the layshaft a smaller gear meshes with a large gear on the output shaft, driving the output shaft clockwise and even more slowly.

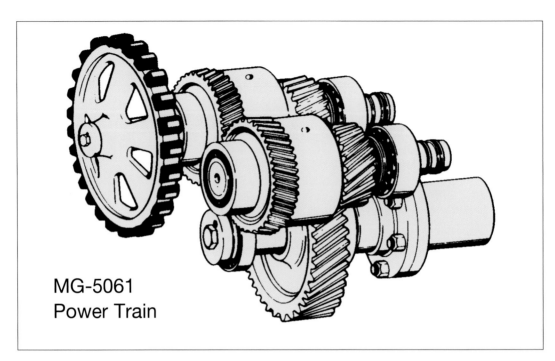

MG-5061
Power Train

Fig 23 Layshaft gearbox.

Astern gear is achieved by a second layshaft. It's very similar to the first, with two gearwheels, one of which has a clutch that can leave it free-spinning or lock it on to the shaft, while the other is constantly in mesh with the output shaft. The key feature about the second layshaft is that it is not driven directly by the input shaft, but by the first layshaft. This means that it's turning in the opposite direction – so if one clutch is released and the other engaged, the direction of the output shaft will be reversed.

EPICYCLIC GEARBOXES

The third alternative – renowned for its smooth operation, but now becoming less popular – is known as the *epicyclic gearbox* (Fig 24).

The central gearwheel – known as the *sun gear* – is mounted on the input shaft, driven by the engine. Around it are several smaller *planet*

gears mounted in a rigid cage called the *planet holder*, while the whole assembly is surrounded by a cylindrical *ring gear* that looks rather like the hub of a trailer wheel but with teeth machined into its inner surface. The output shaft is coupled directly to the planet holder.

For ahead operation, a clutch is used to lock the planet gears so that they can't turn. In effect, this means that the input shaft, ring gear, planet holder and output shaft are all locked together so the whole assembly turns as though it were one piece.

For astern gear, the ring gear is locked to the casing, and the planet gears are left free to rotate. If the input shaft turns clockwise, this means that the inner planets have to turn anticlockwise, driving the outer planets clockwise. In order to do this, they have to run anticlockwise around the inside of the ring gear, taking the planet holder – and, of course, the output shaft – around with them.

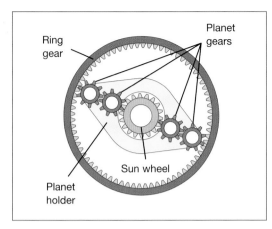

Fig 24 Epicyclic gearbox.

One major snag with this is that although it provides an ahead/astern facility, it can't offer the reduction in shaft speed that is one of the main reasons for wanting a gearbox in the first place. To do that requires a second gearbox in tandem with the first – adding to the cost, weight, bulk and complexity of the whole unit. The second gearbox is usually another epicyclic unit, but with the important difference that there are no clutches involved, and the output shaft is connected to the ring gear instead of to the planet holder.

In this case, the spinning input shaft coming from the reversing gearbox drives the sun wheel, which makes the planet wheels turn, and they, in turn, drive the ring gear.

CLUTCHES

An important characteristic of all marine gearboxes is that the gears of both drive trains – ahead and astern – are in mesh all the time, and that selection of ahead, astern and neutral is achieved by using clutches to disconnect whichever drive train is not in use.

There are two main types of clutches: the *cone clutch* used in the MS2, and the *plate clutch* used in most of the others.

The plate clutch consists of a stack of flat friction plates. Some of the plates have internal splines which mesh with matching splines on the shaft, so they have to turn with it. Sandwiched between them are similar plates which are free to spin around the input shaft but have external splines that mesh with splines on the inner surface of the clutch casing – connected to the output shaft. So long as the stack is loose, the two shafts are free to rotate independently of each other. Compressing the stack forces the plates into contact with each other so that they grip and force the two shafts to turn together.

For relatively small engines – up to about 100 hp – a hand-operated lever may be enough to operate either sort of clutch, but bigger engines need more pressure to be applied. This calls for hydraulics, in which a pump built into the gearbox creates hydraulic pressure to force the plates together. All the gear lever has to do is operate a valve, diverting the hydraulic pressure to whichever clutch requires it.

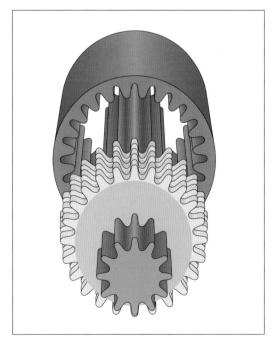

Fig 25 Plate clutch.

... THINGS TO DO

The additives that make modern oils better for your gearbox make them worse for you. Take care to avoid unnecessary or prolonged contact with gearbox oil – new or used.

Only a few teeth of each gearwheel are in mesh with each other at once, and yet the very small surfaces in contact have to transmit the entire power of the engine. When you think about a gearbox in this way, it becomes obvious that the loads involved are very large indeed, and that good lubrication is essential.

Leaks aren't always obvious, especially if they occur between the engine and gearbox, but an external visual inspection will often reveal minor gearbox faults before they develop into major problems. Look out for loose mounting bolts or control cables, as well as for oil leaks around the shaft, hoses and oil cooler.

a. If the gearbox shares the engine's oil, regular oil changes are essential: the acidic by-products of combustion in old engine oil will cause corrosion and rapid wear in the gearbox.

b. If the gearbox uses its own oil, the level should be checked regularly – ideally every day the engine is used. Some types need the engine to be run before the oil level is checked: consult the manufacturer's handbook for this, and to find out whether the dipstick has to be screwed in to show the correct level. Gearboxes don't 'use' oil as an engine does, so if the oil level falls it is almost certainly due to leakage.

c. Having dipped the oil, top up if necessary, usually through the dipstick hole or through a separate filler cap.

Different types of gearbox need different types of oil AND THEY DON'T MIX! Check with the manufacturer's handbook to find out whether your gearbox uses monograde engine oil, gear oil or automatic transmission fluid, and make sure you keep the right stuff on board.

Gearbox dipsticks are often rather awkward to get at, but it's important to make sure that you put it back exactly the way it came out, or it may come into contact with one of the spinning gears. Ensure that it is pushed fully home or screwed right in.

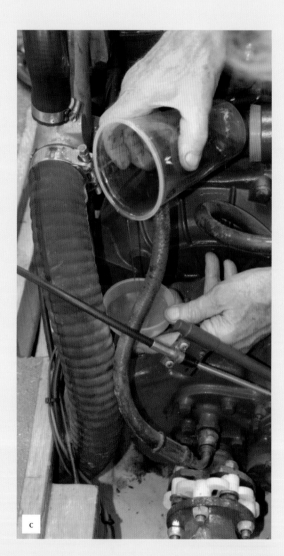

c

PROPELLER AND STERN GLANDS

When you're concentrating on the engine itself, it's easy to forget that the object of the whole thing is to turn the propeller. Whilst it's true that there is little that can go wrong with a propeller, apart from physical damage such as bent, chipped or broken blades, it's worth being sure that you've got the right one for the job; a propeller needs to be carefully matched to the boat, engine and gearbox.

The science of propellers is remarkably complicated, but you can think of a propeller in any of three ways:

♦ as a screw

♦ as a pump, or

♦ as a foil.

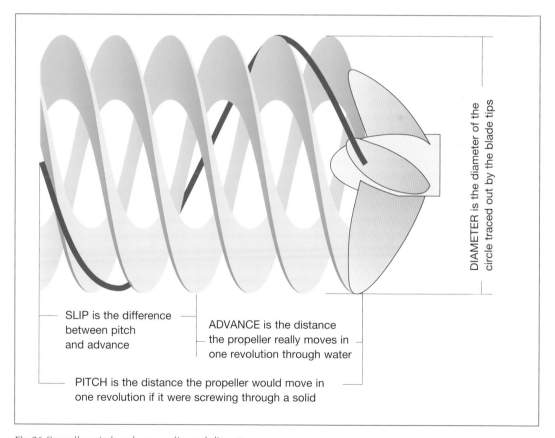

SLIP is the difference between pitch and advance

ADVANCE is the distance the propeller really moves in one revolution through water

DIAMETER is the diameter of the circle traced out by the blade tips

PITCH is the distance the propeller would move in one revolution if it were screwing through a solid

Fig 26 Propeller pitch, advance, slip and diameter.

None of them tells the whole story, but between them they provide a working knowledge of a subject which could easily fill several books, each much bigger than this one.

THE PROPELLER AS A SCREW

One theory regards the propeller as a screw, winding its way through the water like a bolt winding its way into a nut. It's not a very good theory, because a propeller doesn't work in a solid medium, but it explains some of the terminology.

Pitch is the distance the propeller would move in one revolution if it were screwing its way through a solid. In practice it does not move quite that far: the distance it really moves is called the *advance*. The difference between pitch and advance is called *slip*.

The picture is complicated by the fact that the advance isn't quite the same as the distance the boat moves: for one thing the boat tends to drag some water along with it, and for another, the water directly affected by the propeller tends to move astern compared with the water nearby. This makes it very difficult to measure the true slip. It's very much easier to find the *apparent slip*, which is the difference between the boat's movement and the pitch. For high-performance motorboats, the apparent slip may be as little as 10 per cent of the pitch, but for a heavy motorsailer or workboat it can be as much as 50 per cent.

THE PROPELLER AS A PUMP

Another theory treats the propeller as a pump, squirting water backwards. Newton's third law of motion says that for every action there is an equal but opposite reaction, so if the propeller pushes a lump of water backwards, the water

also pushes the propeller (together with the boat to which it is attached) forwards.

The amount of water the propeller can push in a given time depends mainly on the power available: a 100 hp engine, for instance, can move about 50 ft^3 per second. You can think of that 50 ft^3 of water as being in the shape of a cylinder, whose diameter is roughly the same as that of the propeller, and whose length is roughly the distance the prop has moved through the water in the time. For a slow motor cruiser, moving at 10 ft/sec, that means the cylinder must have a cross-section of 5 ft^2, or a diameter of about 30 in. For a sportsboat doing 50 ft/sec with the same engine, the cylinder's cross-section must be reduced to 1 ft^2, or about 13 in.

This suggests – correctly – that the diameter of a propeller should vary depending on the power transmitted (because that determines the volume of water it can move in one second) and the speed at which the boat is moving (because that determines the length of the cylinder).

THE PROPELLER AS A FOIL

A more recent, realistic and complicated theory regards each blade of the propeller as a foil, like a boat's sail or an aircraft's wing.

Like a sail, the blade has to be at a slight angle to the fluid flowing over it if it is to generate any useful force. Unfortunately, increasing this *angle of attack* doesn't just increase the useful thrust: it also increases the drag – which is one of the reasons why over-sheeting a sail makes it less efficient.

For a sail, the optimum angle between the sail and the air flow is about 20–25°, but for a propeller blade in water it is much smaller – about 4°.

The analogy with sailing goes on if you think about how the flow across the propeller blade is created. The 'apparent wind' flowing across a sail is made up of two components: the 'true wind'

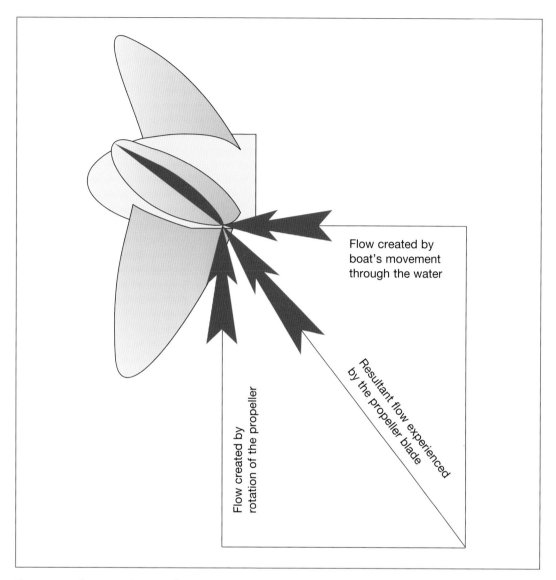

Fig 27 Water flow over the propeller.

that would be felt if the boat were stationary, and the 'induced wind' caused by its own movement through the air. In the case of a propeller blade, the equivalent of the true wind is created by the rotation of the propeller, while its 'induced wind' is caused by its movement through the water.

That is why a propeller blade needs to be twisted: 'induced wind' is much the same all over the propeller, but the 'true wind' varies

dramatically, because the tip of each blade sweeps around a much bigger circle than the sections of the blade that are nearer the shaft.

It also suggests that the pitch of the propeller needs to be carefully matched to the speed at which the water is moving through the propeller: a fast boat generally needs much more pitch than a slow one, in order to cope with the much greater 'induced wind'.

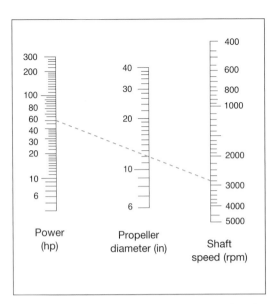

Fig 28a Propeller selection diagram. Lay a straight edge across the two known variables to find the unknown.

Example: For 50 hp delivered through a shaft turning at 2,800 rpm, the optimum diameter is approximately 12 in.

CHOOSING A PROPELLER

What all this boils down to is that the choice of propeller depends on a mass of interrelated variables that include:

♦ the power available
♦ the shaft speed, and
♦ the speed of the propeller through the water.

There are so many variables (even the temperature and salinity of the water have a part to play) that trying to work out the pitch and diameter of the ideal propeller for a particular boat from pure theory is almost impossible: it is invariably better to leave it to a specialist such as a propeller manufacturer who will have access to a mass of experience and information, and can probably come up with the right answer simply by tapping a few vital statistics into a computer. It's important to do so, because the wrong

propeller can easily make a big difference to the boat's performance.

If you want to do a quick check, however, the diagrams (Fig 28a and 28b) provide a reasonably accurate estimate of the appropriate diameter and pitch.

CAVITATION AND VENTILATION

Although the words 'cavitation' and 'ventilation' are often used interchangeably, they are really two quite different things. What they have in common is that they can rob a propeller of almost all its grip on the water.

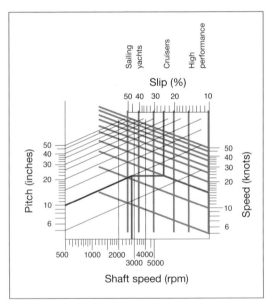

Fig 28b Propeller pitch diagram. Estimate a likely value for slip. From that figure on the top scale, drop vertically to meet the diagonal corresponding to your estimated speed.

From that intersection, draw a horizontal line to meet the vertical drawn upwards from the shaft speed on the bottom scale. Where these two meet gives the optimum pitch.

Example: For a boat expected to achieve 15 knots with a 50 hp engine, and a shaft speed of 2,800 rpm, the optimum pitch is approximately 10 in.

Ventilation is most common on sports boats and on sailing boats with outboard motors mounted high on their transoms, because it is caused by air being sucked into the propeller. A propeller designed to operate in water obviously can't work very well in air: it generates very little thrust and suffers very little drag, so the engine speed increases while the boat slows down.

Cavitation is caused by the propeller itself. As it rotates, it creates thrust by increasing the pressure on the aft faces of its blades and simultaneously reducing the pressure on their forward faces. As the pressure falls, the boiling point of water reduces, eventually reaching the stage at which it will boil, even at sea temperature. When this happens, bubbles of water vapour form on the forward face of the propeller. The immediate effect of severe cavitation is very much like ventilation: the engine races but the propeller ceases to generate much thrust. The long-term effect of even minor cavitation is an erosion of the propeller blades known as 'cavitation burn', caused by the collapsing bubbles.

Cavitation can be caused by using the wrong propeller for the boat, such as one with too much pitch or not enough blade area, but can also be caused by damage to the propeller, such as nicks in the leading edge of the blade.

STERN GLANDS

It may be stating the obvious to say that a propeller works best in water, and that an engine (generally) works in the air inside the boat, but it's an important point because it means that somehow the shaft that turns the propeller has to pass out through a hole in the hull without letting water in. The component responsible is a *stern gland.*

Engineers have faced this problem for centuries, and one can imagine the joy of some Egyptian farmer when he discovered that his slave-powered irrigation pump worked better if he stuffed reeds around the handle to stop it leaking, or the relief felt by a Viking longship's crew when they found they got less wet if they used rags to fill the gaps between their oars and the row of shields.

Those simple discoveries were the direct ancestors of what is still the commonest kind of stern gland – the *'stuffing box'.*

STUFFING BOXES

A stuffing box is basically an extension of the boat's stern tube – the tube that carries the propeller shaft out through the hull – enlarged to form an outer casing that is filled with rings of grease-soaked packing. A sleeve around the shaft holds the packing in place, and can be tightened down to compress the packing so that it forms an almost watertight seal. A common refinement – essential for engines mounted on modern 'soft mounts' – is a short length of flexible tube between the stern tube and the stuffing box, to allow the stuffing box to move with the shaft as it vibrates.

Another is a reservoir of grease, so that the packing can be re-greased without having to dismantle the whole thing. Obviously this only works if you use it, so it makes sense to get into the habit of tightening the greaser every time the engine is used.

Even with a good supply of fresh grease, an inevitable problem with a traditional stuffing box is that if the stuffing is compressed tightly enough to make a perfect seal, there would be so much friction that the packing would get hot – drying out the grease and wearing away both the packing and the shaft. For this reason, a stuffing box should never be tightened down so much that it is completely dry: a slight leak helps cool and lubricate it without imposing too much strain on the bilge pump. This is so important that on planing boats, where the movement of the boat itself might suck water away

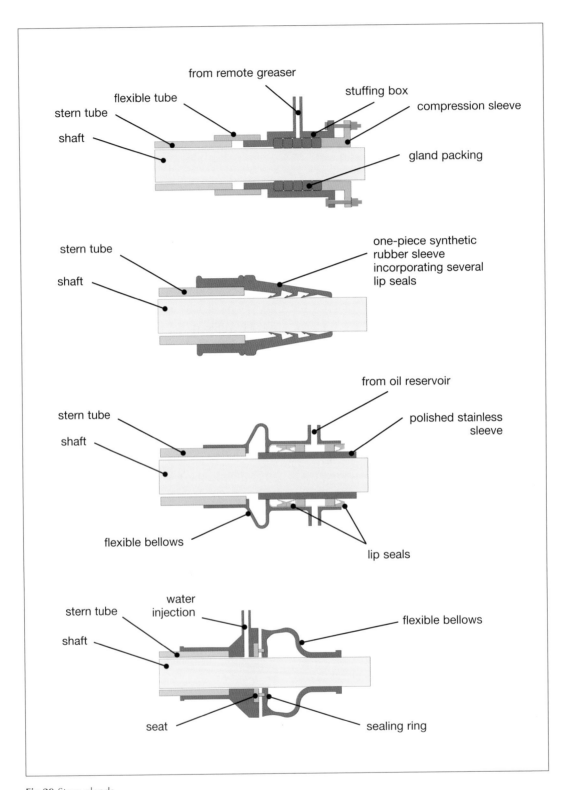

Fig 29 Stern glands.

from the stern gland, water is often pumped into the stuffing box by the cooling system.

OTHER SHAFT SEALS

This perceived 'problem' of essential leakage – and the rather fiddly job of re-packing – has led to the development of several more modern alternatives.

Some use a synthetic rubber diaphragm clamped to the shaft or stern tube to hold a ceramic or carbon sealing ring in firm contact against a phosphor bronze or stainless-steel seat. The virtue of this is that there's nothing rubbing against the shaft, and no need to mess about with grease: the only maintenance required is occasionally to pull the diaphragm away from the seat, to flush out any debris and stop the two parts sticking together if the boat is left idle for any length of time.

Another consists of a hollow rubber cone with a row of *lip seals* – similar to those that prevent oil escaping from the engine itself – moulded into its inner surface. These need greasing once a season or every 200 hours, and have to be 'vented' to ensure that they are full of water each time the boat is launched.

A third variety keeps water out by surrounding the shaft by a tube filled with oil. Lip seals keep the oil inside the tube, by sealing against a polished stainless-steel sleeve fitted to the shaft.

OUTDRIVES AND SAILDRIVES

Outdrives and saildrives circumvent the need for a stern gland altogether by – in effect – extending the gearbox right into the water in the form of a drive leg. Of course, this doesn't make the problem go away altogether: it just changes it a bit, because instead of having to seal the gap around a small rotating shaft, there's a much bigger hole to plug between the hull and the drive leg. Outdrives achieve this by using flexible rubber bellows between the transom and the leg, while saildrives use a rubber diaphragm to seal the joint between the drive unit and the bottom of the boat. Servicing recommendations vary depending on the manufacturer and model, but the importance of following them to the letter is obvious.

... THINGS TO DO

PROPELLERS

It is a good idea to remove the propeller from time to time, to stop it seizing solid on the shaft. This is particularly important in the case of outboard and outdrive propellers, which usually have either shear pins or flexible hubs that are designed to protect the engine and transmission by slipping or breaking if the propeller hits something. Unfortunately, after a period of time, these protective devices break of their own accord, so it's important to make sure that the propeller can be taken off and replaced at sea.

REMOVING A PROPELLER

a. Straighten out the legs of the split pin and remove it, or flatten the lugs of the tab washer that holds the propeller nut in place.

b. Use a block of wood to stop the propeller turning, and undo the propeller nut.

c. It should be possible to pull an outboard or sterndrive's propeller off the shaft by hand. Make a note of the order that any springs or washers come off, and which way round they go.

d. The best way to remove a stubborn propeller is with a puller. Once the puller has been tightened, a few light taps with a hammer on the puller should make the propeller slide off.

e. As a last resort, try warming the propeller boss with a blow lamp.

REPAIRING A PROPELLER

If necessary, use a file to smooth any burrs or nicks from the edges of the propeller, but be careful not to remove so much metal that there is a risk of the propeller becoming unbalanced. If it is badly damaged, refer it to a propeller repair workshop.

REPLACING A PROPELLER

a. Before replacing the propeller, grease the shaft with a non-graphite grease.

b. Slide the propeller back on, making sure that any keys, shear pins, springs or washers are replaced exactly as they were removed, but use a new tab washer (if there was one) to hold the propeller nut in place.

c. Tighten the propeller nut just sufficiently to grip the propeller, then insert the split pin (if there was one). Bend up one tab of the tab washer to hold the nut in place, or open the legs of the split pin and bend them round the nut.

STERN GLANDS

1 GREASING A STERN GLAND

If your stern gland is fitted with a greaser, routine greasing involves either turning the handle or tightening the cap by one turn after every few hours' running: once every four hours is about right. Eventually, this will use up all the grease, which will have to be replaced. Don't use graphite grease: its carbon content causes extremely rapid corrosion in the presence of salt water.

a. Remove the top cap of the greaser. If yours is the type that uses a handle like that of a garden tap to operate a plunger, 'unscrew' the handle to return the plunger to its starting position.

b. Invert the grease can, so that the hole in the plastic disc is over the greaser.

c. Press the whole can downwards, so that the plastic disc slides into the can and forces grease out. You may have to stop a few times, especially when the greaser is nearly full, to allow trapped air to escape.

d. Replace the cap of the greaser.

2 ADJUSTING A STERN GLAND

a. With the engine stopped and out of gear, turn the shaft by hand to get a feel for how stiff it is.

b. Slacken off any lock-nuts holding the two parts of the stern gland together. Tighten the clamping bolts that hold the two parts of the stern gland together, or screw the sleeve into the outer casing. Don't tighten them by more than half a turn at a time, and – if yours is the type that has two or three bolts – make sure that you tighten them evenly.

c. Turn the shaft by hand and check for drips: it should turn freely and leak between one and six drips per minute. If it is leaking too quickly, tighten the clamping bolts further.

2a

2b

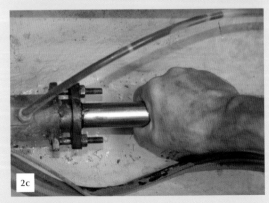
2c

3 RE-PACKING A STERN GLAND

If the stern gland cannot be tightened enough to reduce the leakage to an acceptable level without making the shaft unduly stiff to turn, then the packing needs to be replaced. **This can only be done with the boat out of the water.** (Photos 3a–g show the stern gland viewed from above.)

a. Remove the lock-nuts and nuts that hold the two parts together, or unscrew the two parts completely. If necessary, use two large screwdrivers or pinch-bars to prise the two parts apart.

b. Remove the old packing. This will probably involve fashioning some home-made 'tools', such as a small screwdriver filed to a point, or bent coat-hanger wire.

c. Use a degreasing agent (such as 'Jizer' or 'Gunk') to clean the shaft and stern gland.

d. Check that the new packing is the right size: it should just fit the gap between the shaft and the outer casing of the stern gland.

e. Wrap the packing around the shaft, making sure that it sits squarely on the shaft, and then use a sharp craft knife to cut through the packing along the line of the shaft so as to produce several short lengths of packing, each just long enough to fit round the shaft. Use the stern gland's end cover to push each ring of packing down into the stern gland case, making sure that each one sits squarely on the shaft and that its cut ends are not lined up with the cut ends of the one before.

f. Reassemble the end cover and clamping bolts, but leave them finger-tight. Turn the shaft several times by hand in order to help bed the packing rings together.

g. As soon as the boat is re-launched, adjust the stern gland to the required slow drip.

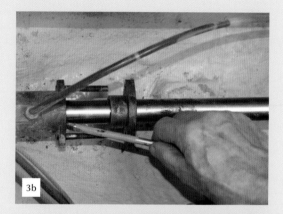

3b

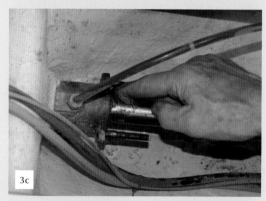

3c

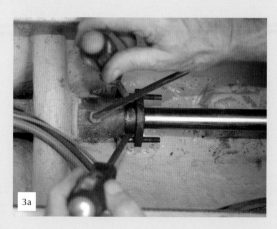

3a

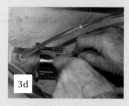

3d

3e

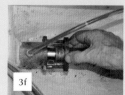

3f

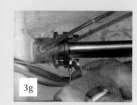

3g

CONTROL SYSTEMS

10

A few small open workboats still have engine and gearbox controls that can be operated directly by the helmsman, but that simple system is becoming increasingly rare on pleasure craft, where it is very much more common to find some kind of remote control system.

As is often the case on boats, there is no 'standard' arrangement: push-pull cables are by far the commonest, but even these are available in several different forms and face competition from hydraulic systems – in which the remote control lever operates a pump connected by pipes to a hydraulic ram that operates the engine controls – and electronic systems that use wires to carry control signals from the wheelhouse to the engine room.

CABLE SYSTEMS

The cables that are used in most systems are rather like those used to work the brakes on a bicycle, with a central control cable inside a tubular outer casing. The casing is fixed at both ends, so that when you pull one end of the inner cable, the other end retracts.

Boats' control cables are usually very much bigger and more robust than those on a push-bike, but the main difference is that instead of using very flexible multi-strand wire for the inner cable, marine systems use a single strand of stiff wire that enables them to push as well as pull.

CONTROL HEADS

It's easy to see how cables can be used in a *twin-lever* system, where one lever controls the engine and another operates the gearbox. Pushing the top of the 'throttle' lever forwards, for instance, pulls on the cable, which in turn pulls the lever on the engine's fuel pump.

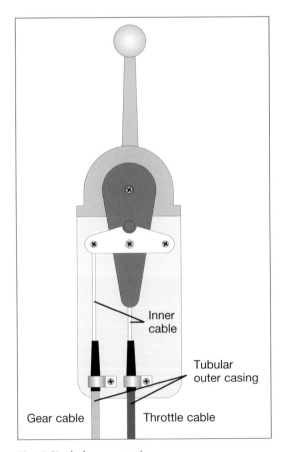

Fig 30 Single-lever control.

Single-lever systems, in which engine and gearbox are controlled by the same lever, are generally more popular – but are more complicated because they have to achieve a positive gear shift between ahead, neutral and astern, but also offer progressive control of the engine speed.

This is achieved by connecting the 'throttle' cable directly to the control lever, while the gear cable is connected to a horizontal seesaw arrangement. A peg sticking out of the throttle lever engages in a notch in the top edge of the seesaw, so that the first few degrees of movement of the control lever is enough to rock the seesaw so that it pulls or pushes on the gear cable. The geometry of the arrangement is such that this first movement of the control lever has virtually no effect on the throttle cable. Once the seesaw has been rocked far enough to engage gear, however, the peg is clear of the slot, and the lever can move further, pulling on the throttle cable without having any more effect on the gear control linkage.

DUAL STATION CONTROLS

One snag with this system is that it can only be worked with the control lever: pulling or pushing on the other end of the gear cable would have no effect whatsoever. In a boat with only one control station, this is not a problem, but on boats such as motorsailers or flybridge motor cruisers, where the engine may have to be controlled from two different places, it would make the whole system jam up if it were not for a component called a *Dual Station (or DS) unit* that isolates one control head when the other is in use.

A DS unit is based on a sheet-metal frame with two curved slots cut into it. Bridging the gap between the two slots is a metal bar connected to the incoming cables from the two helm positions and the outgoing cable to the engine. The bar is held in place on the base plate by two metal pegs, which pass through the slots. The clever bit about this arrangement is that the pegs are slightly further apart than the

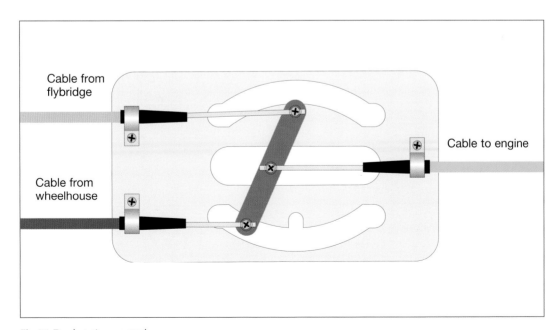

Cable from flybridge

Cable from wheelhouse

Cable to engine

Fig 31 Dual station control.

slots: the only reason the pegs can fit into both slots at once is that each slot has a semicircular cut-out. When one peg is nestling in its cut-out, the other peg is free to slide along the slot and vice versa.

If, on a motorsailer, you move the saloon control lever while the cockpit control is in neutral, the cockpit cable's peg drops into its cut-out, allowing the saloon cable to move its end of the bar to push or pull the outgoing cable. When the saloon control is in neutral, any movement of the cockpit control is enough to move the bar so that its 'saloon' end drops into its cut-out, leaving the 'cockpit' end free to move – with the same effect on the outgoing cable.

CABLES

Cable systems are generally reliable so long as they are properly installed in the first place and then receive a certain minimal level of maintenance. It's worth doing, because if your control system fails, you will almost certainly look foolish (bystanders never believe that a messed-up manoeuvre was caused by mechanical problems), and you may well face serious damage or injury.

When problems do occur, control failure is more often due to problems with the cable, rather than to the control units themselves:

♦ Internal corrosion, caused by water getting in through splits in the outer cable, can jam the inner cable.

♦ Wear on the inner surfaces of the outer cable allows the inner cable to slop from side to side, producing excessive backlash (free play) on the cable end: on a gear cable this may mean that there is not enough controlled movement to operate the gear lever properly.

♦ Bent or corroded end rods make operation stiff and may eventually lead to the inner cable stretching: again, this is particularly serious in the gear cable.

♦ Worn, corroded or disconnected end fittings. The split pins that secure the cable to the gearbox and fuel pump are so thin that they are particularly prone to corrosion, but look out too for the clamping arrangements that hold the outer cable in place.

♦ Poorly designed cable runs can make controls stiff from the outset, and give rise to a lot of backlash. Over time, this gets worse, as the bends cause increased wear and tear inside the cable: ideally cables should be dead straight, with no long sweeping bends or bends tighter than about 8 in (20 cm) radius.

... THINGS TO DO

a. Periodically – about once a season, but more often in exposed locations – inspect the cable end fittings for wear and corrosion. Replace split pins with new ones if they are corroded, or if they have been removed for any reason.

b. With the inner cable in its fully extended position, lightly grease the exposed part with a non-graphite grease.

c. Clean and re-grease the moving parts of control heads and DS units. Inspect the cable run, looking for splits or wear in the outer plastic sheath – often given away by rust streaks.

a

b

c

TOOLS AND WORKING PRACTICES

11

A lot of work on an engine involves removing and replacing components, so it involves dealing with a wide variety of fastenings – literally, 'getting down to the nuts and bolts'.

Bolts vary in length and diameter, but come in a number of standard sizes quoted in imperial or metric measurements. Metric measurements are now used almost universally on British, European and Japanese equipment, while imperial sizes are found on old British equipment and on almost anything intended primarily for the American market. This means that you are unlikely to come across a mixture of the two on one engine, but doesn't mean to say that you won't find imperial fastenings on the engine and metric elsewhere on the boat – or vice versa.

Even if you find two bolts of exactly the same diameter, they may not be interchangeable because there are a variety of different 'standard' screw threads which differ in cross-section as well as in the number of threads per inch (Fig 32). It's important to make sure that you match the right nut and bolt together and that you screw bolts or studs back into the holes they came out of, because although some odd combinations are compatible, the vast majority are not. Unless a nut or bolt is clearly in poor condition, it should turn smoothly and easily until it reaches the final tightening-up stage: if it starts easily but suddenly becomes stiff for no obvious reason, or if it feels unusually floppy, it's a pretty safe bet that it's the wrong one for the job.

If the differences between screw threads seem subtle, the differences between their heads certainly are not: it is obvious that you can't use a spanner to undo a bolt with a domed and slotted head intended for a screwdriver! It's surprising, though, how often DIY mechanics find themselves trying to work with spanners or screwdrivers that don't quite fit. It's important to use the right tool for the job, because although a 13 mm spanner will just about cope with a 1/2 in hexagon head, it is slightly too large. The difference is only about a quarter of a millimetre, but that is enough to allow the nut to twist between the jaws. Instead of the load

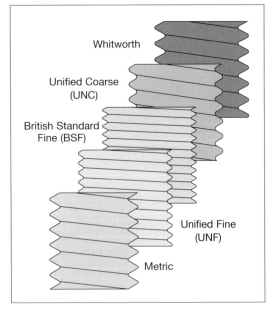

Fig 32 Screw thread sizes.

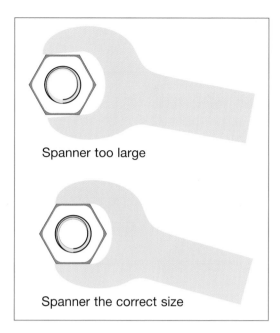

Fig 33 Spanner sizes.

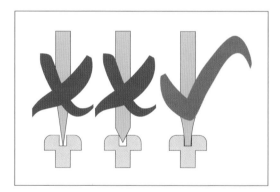

Fig 34 Screwdriver shapes.

being spread across the flats of the nut, it's then concentrated at the corners, so as soon as you try to apply any appreciable force to the spanner, the nut levers the jaws of the spanner apart, while the spanner rounds off the corners of the nut.

Similar comments apply to screwdrivers – probably the most misused tools of all. Not only do you need the right kind of screwdriver – flat-bladed for slotted heads, and cross-point for cross-heads – but it should also be the right size

for the job. A screwdriver that is too big won't go into the slot at all, but one that is too small will tend to twist out of shape and damage the edges of the slot. A flat-bladed screwdriver should be ground to a fairly shallow taper, so that it doesn't try to lever itself out of the slot, and have a sharp-edged square tip so that it doesn't hit the bottom of the slot before it makes contact with the sides.

The drive towards more compact engines has made engine manufacturers turn increasingly to fastenings with 'unconventional' heads. *Allen screws*, in which the head of the bolt has a hexagonal recess, are now so common that they hardly rank as unconventional. Dealing with these requires either a screwdriver equipped with a selection of appropriate 'bits' or a set of Allen keys – bent pieces of hexagonal hardened steel bar. You may need lots of them: Allen screws come in a variety of metric and imperial sizes, and a good fit between the tool and the fastening is even more important than for a spanner – not least because if you damage the head of an Allen screw it can be exceedingly difficult to remove.

PIPE FITTINGS

Pipe fittings also come in a variety of shapes and sizes, designed to cope with pipes of different materials and different sizes, and with operating pressures ranging from partial vacuum to several thousand pounds per square inch. There are, however, only three main types that you are likely to come across.

Pipe clips are bands of thin metal designed to fit round flexible pipes. A bolt or screw joins the ends of the band so that it can be tightened to compress the pipe against the rigid spigot to which it is attached. They're found in all sorts of relatively low-pressure applications, from marine toilets to engine cooling systems and exhaust pipes.

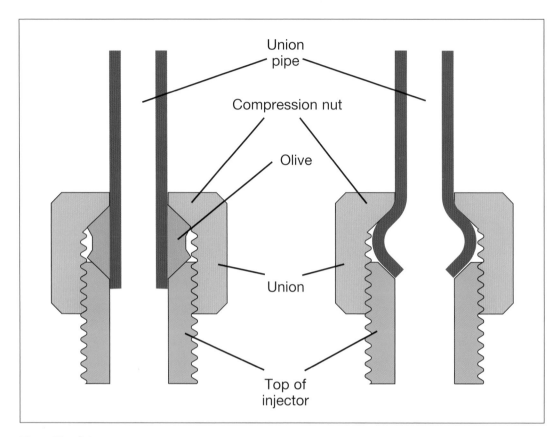

Fig 35 Pipe fittings.

Removing a hose clip is a simple matter of unscrewing the clip until it is loose enough to slide along the hose and then pulling the hose off the spigot. In practice, a common problem is that the hose may have glued itself to the spigot, in which case it may come free if you massage the hose to loosen the bond, and then prise it off with a screwdriver. As a last resort, a flexible pipe can always be cut, but do make sure you have a replacement available before you do so.

Replacing a hose can be more difficult, because it may be such a tight fit on the spigot that you wonder why it needs a hose clip at all. Dipping the end of the hose in boiling water may help by softening it, and a smear of washing-up liquid can be used to provide some gentle lubrication. Whatever method you use, it's

worth threading the pipe clip on to the hose first, so that you don't have to unscrew it completely in order to fit it on to the pipe in situ.

Fitting the pipe on to the spigot is especially difficult if the spigot has a bulge or ridge around its end, but these things are not put there just to make life awkward: they are intended to provide extra security once the pipe is in place. They only work, however, if you make sure that the hose clip is on the right side of the ridge – nearest the root of the spigot so that the pipe would have to drag the pipe clip over the ridge in order to pull itself free.

Compression fittings are used on rigid pipes, or occasionally on flexible pipes with a rigid insert. There are two types, but both look like an unusually deep nut with the pipe sticking out of the middle.

One kind, used mainly on relatively low-pressure applications such as domestic plumbing and sometimes on the low-pressure side of the fuel system, uses a straight-ended pipe with a brass or plastic ring called an *olive* threaded on to it. The end of the pipe fits into a recess in the spigot, but the olive rests on top. Then, when the nut is tightened down on to the threaded spigot, the olive is compressed between the 'nut' and the spigot, to grip the pipe and form the seal.

The only cause for concern when working with this type of pipe fitting is that over-tightening it can distort or split the olive.

High pressure pipes, such as injection pipes, use a development of this principle in which the pipe itself is shaped to form the olive. Making the joint in the first place requires special equipment, but once the pipe has been shaped it forms a secure and leak-proof joint that is as easy to do up or undo as a nut and bolt. Perhaps the biggest danger is the temptation to treat these kinds of joint as though they are flexible: they're not, so don't try to bend or move the pipe without slackening the unions that hold it in place.

The third kind of pipe union is called a *banjo bolt* – so-called because one part of it is supposed to look like a banjo. The 'banjo' is a metal fitting that has to be brazed or soldered on to the pipe, and which then forms a hollow channel right around the central bolt. The 'bolt' part looks exactly like a conventional bolt, except that its head is often rather larger and thinner than you might expect, and its shaft is a hollow. A hole in the side of the bolt allows liquid to flow from the pipe, around the circular channel formed by the banjo, and down through the hollow bolt. Brass, fibre or nylon washers form a seal around the upper and lower edges of the banjo to stop leaks: be careful not to lose them when undoing a banjo bolt, and make sure that they go back when you replace it.

SEALS AND GASKETS

Other joints in an engine, such as those between the face plate and casing of a water pump or between the injectors and the cylinder head, are just as important as the joints in the 'plumbing' that surrounds it. It's difficult to get a good metal-to-metal seal over a wide area, especially as the parts concerned may be expanding and contracting at different rates when the engine warms up and cools down. To overcome the problem, these kinds of joint usually include a *gasket* of more resilient material.

Some gaskets, especially cylinder head gaskets, are quite sophisticated components in their own right, including layers of different materials and inserts to withstand particular stresses or to help them stay in shape, but the majority of gaskets are relatively simple sheets of steel, copper, cork, rubber or paper.

Copper gaskets are used in small, highly stressed joints such as between an injector and its seat, and usually look much like an ordinary washer. Ideally, a copper gasket should only be used once, because it loses its resilience once it has been fitted. If necessary, though, it can be renovated by heating it over a gas ring until it turns blue, and then immediately dropping it into cold water.

At the other extreme are *cork* or *rubber gaskets* used in places such as the joint between the rocker box cover and the cylinder head. There are no great stresses involved here, but the rocker box cover is so thin that it would easily distort if it had to be bolted down very firmly to achieve a seal. The only snag with these is that the gasket almost invariably sticks to one or the other of the two components: if part of it sticks to one and the rest of it sticks to the other, you may need to peel it away very carefully with a thin knife to make sure it comes away in one piece.

Paper gaskets are less resilient than cork or rubber, but they work well and are so cheap that although it's almost impossible to salvage one once it's been used, there's no excuse for not using a new one every time. If you can't get a ready-made replacement, it's easy to make your own from a sheet of gasket paper. In a real emergency, it's even possible to use writing paper or a breakfast cereal packet as a stopgap!

The first step in the process of replacing a gasket is to remove all traces of the old gasket from the surfaces. You may need to use a scraper, but be careful not to scratch the machined surfaces – if you must use a screwdriver for the job, file the corners off first! Then, if you're making a new paper gasket, smear one of the mating surfaces with a little oil, and press it down hard on to the gasket paper so as to leave a clear picture of its shape and any holes that are needed.

Once you've cut out the new gasket, use another smear of oil to stick it temporarily to one surface, and make future separation of the joint easier. Then reassemble the joint and tighten all the bolts evenly.

TOOLS

The standard advice about tools is 'buy the best you can afford'. All too often that advice goes with a list of 'essential' tools and spares that would not disgrace a professional workshop, but would go a long way towards sinking a small cruising yacht!

There's no doubt that high-quality, high-price tools such as those made by Britool, Gedore and Snap-On are virtually indestructible and a pleasure to use, but unfortunately they sink just as quickly as any others if they get dropped overboard or in the bilges, rust nearly as quickly once they've been exposed to salt water, and are exceptionally prone to being 'borrowed'.

Very cheap tools, such as many of those imported from India and China, will let you believe you've got a full tool kit until you try unscrewing a tough nut in an awkward spot. That's when you find out that the reason they are cheaper is because they don't fit as well in the first place, and that they get worse once they start to bend under the strain.

Fortunately there are plenty of mid-range tools made by companies such as Stanley and Kamasa, which are nearly as good as the front-runners but are a fraction of the price.

TOOL KIT

A set of *open-ended spanners*, of the right sizes – imperial (AF) or metric – is essential, and if you have a very old boat or engine, you may need the now-obsolete Whitworth sizes, too.

Ring spanners are better if you need to use much force or work in an awkward position, but can't be used on pipe fittings or lock-nuts. As you need two spanners of the same size to undo a nut and bolt anyway, it makes sense to have one set of rings and one of open-ended.

Combination spanners have a ring on one end and open-ended jaws on the other, but as both ends are usually the same size, you will still need two sets!

Socket spanners make life much easier, and are the only way of getting at some of the less accessible fastenings on compact modern engines, but it's debatable whether they are essential for basic maintenance. They're nice to have, but no substitute for conventional spanners because there are some fastenings they can't cope with. Whitworth socket spanners are very hard to come by, but a set of tubular box spanners is a reasonable (and economical) alternative.

Few professional mechanics would admit to using *adjustable spanners* on an engine, because they are inevitably less rigid than proper spanners, and more likely to damage the nut or bolt as a result. Every boat, however, has at least one fastening somewhere that is an odd size. When you find out which one it is, a good adjustable will get you out of trouble. Anything less than the best is a waste of space.

Much the same applies to *mole-grips* or *vise-grips*. Few mechanics will admit to using them, but few would be without them!

The most useful *pliers* are the square-ended 'general purpose' type, about 6–8 in (15–20 cm) long. 'Needle-nose' pliers are less versatile, but are a cheap and worthwhile addition to a tool kit, especially if your boat has much electrical equipment on board.

You'll inevitably need several *screwdrivers*, including a couple of cross-point screwdrivers and three or four flat-bladed ones, including a small 'electrician's' and a long-shafted 'heavy duty'. Handle shapes are a matter of personal choice, though the oval handles of 'carpenter's' screwdrivers may allow you to exert more force than the round handles of 'mechanic's' screwdrivers.

Allen keys (or a screwdriver with a wide selection of hexagonal bits) are becoming more and more important.

A *hammer* is a nasty thing to threaten an engine with, but a light ball-pein 'engineer's' hammer is worth having, as is a soft-faced hammer with a weighted nylon head instead of a lump of hardened steel.

Finally, there are cutting tools: a small *hacksaw* and some spare blades; a *craft knife* or large scalpel, and perhaps a small, flat-bladed scraper.

TRICKS OF THE TRADE

The possibility that you might take an engine to pieces and not be able to put it back together again is probably the mechanic's worst nightmare. The best tip for avoiding it is to be scrupulously and relentlessly methodical: lay the bits of your

engine down in the order they were removed, and keep the fastenings with the relevant part – don't tip all the nuts and bolts into one box, because it may then take hours of trial and error to find the right one, and don't complicate matters by taking things apart that you don't have to.

It almost goes without saying that you should refer to the manual for any job that is not completely familiar to you, but if you do find yourself working without a manual, don't be afraid to make notes or sketches of the order in which things came apart or what went where. Look carefully at any component before you remove it, and try to figure out what it does, which bolts hold it on, and which hold something inside it.

Seized fastenings make life difficult, but are a common feature of many boat engines. Before applying brute force, it's important to bear in mind that conventional spanners are designed to apply the right amount of leverage for the fastenings they fit, and that if you lengthen a spanner to unscrew a bolt that is already weakened by corrosion, you may make it shear off completely.

The first stage is to make sure that you are working efficiently:

♦ Try to give yourself as much room and light as you need by removing any covers or hatches that are in the way.

♦ Keep your hands and tools clean so that you can get a good grip.

♦ Pull on the end of a spanner: you are less likely to get hurt if the spanner slips when you are pulling than pushing, and the end gives you more leverage.

♦ Use a ring spanner if possible, rather than an open-ended one.

♦ Turn the nut rather than the bolt.

If that doesn't do the trick, it's worth trying to tighten the offending fastening to break the bond between the threads, and penetrating oil can work wonders so long as it is left alone for long enough to penetrate.

More drastic measures include lengthening the spanner with a length of pipe; and shock treatment by tapping the spanner with a hammer while applying steady pressure by hand.

For a really stubborn fastening, you may have to resort to more destructive methods such as using a cold chisel or a gadget called a nut splitter (like an oversized ring spanner, but with a blade which can be screwed in to break a seized nut) to split the nut, or a hacksaw to cut the side off it. Bolt heads that have rounded off can be carefully filed down to take the next size of spanner, or can have a slot hacksawed in. Most drastic of all, but often quickest, is to cut through the bolt completely.

Destroying a nut and bolt isn't too frightening because it can always be replaced, but the idea of cutting or snapping a stud or bolt that has seized into a casting is more worrying. Even so, it's not the end of the world.

Once the load has been removed from the broken stud, it may unscrew relatively easily, particularly if it has been well soaked with penetrating oil. This is where mole-grips come in handy. Alternatively, you could try cutting a screwdriver slot in the remains, or screwing two nuts down on to it. Once the second nut has been tightened down hard against the first, a spanner applied to the lower nut can be used to unscrew the stud.

If the stud has broken off flush with the casting, it should be possible to drill a hole down the centre of it in order to use a tool called a *stud extractor*. This is a tapered rod of hardened steel with a very coarse thread cut into it. The stud extractor's thread is in the opposite direction to that of a conventional bolt, so as you screw the stud extractor into the stud, it first grips the side of the drilled hole, and then unscrews the stud.

If this doesn't work, it may be possible to drill out the stud altogether, and then use a device called a tap to recut the thread in the hole. Taps, unfortunately, are quite expensive, so it's hardly worth having a full set in hand to cope with the occasional mishap, and it may be worth referring the job to a professional.

The ultimate sanction is definitely a professional's job because of the equipment required, but it's worth knowing that it is possible to drill out the hole oversize, and put in a completely new thread known as a *helicoil insert*.

12

FAULT-FINDING

Even though the individual processes that make a diesel engine run are basically simple, there are so many of them, and so many individual components involved, that trying to trace a particular fault can seem an almost impossible task. It can be done, though!

The first stage is to establish exactly what the symptoms are, and how they developed. Was the engine OK the last time you used it, but playing up now? Have you done anything in between which may have caused the fault to develop? Did it develop suddenly, or gradually?

Armed with this information, the daunting-looking fault-finding lists that follow can be whittled down to a more manageable size.

Suppose, for instance, that we're faced with an engine overheating. The fault-finding list offers a wide range of possibilities, ranging from a faulty gauge to wrong valve timing. Working through all of them could easily take a professional mechanic a full day or more, so it's important to establish what is *likely*, as opposed to what is *possible*.

Assuming you haven't done anything to change the valve timing or fuel pump timing, it's most unlikely to be either of those. Suppose, however, that you shut the seacock in order to check the raw-water filter, and the engine overheated quickly the next time you started it. In that case, it's a fair bet that you may have forgotten to open the seacock again!

Having narrowed down the possibilities, it's then time to work through them in a logical order. Try to avoid making random changes or adjustments to your engine: don't do anything without a good reason, and always remember (or write down) what you've done so that you can undo it if it doesn't solve the problem or makes it worse.

Most important of all, though, remember that common, simple problems, such as a blocked fuel filter or leaky pipe, are much more likely than more obscure or complicated ones such as a sheared fuel pump drive.

STARTING PROBLEMS

THE STARTER WON'T TURN THE ENGINE FAST ENOUGH TO START ... OR AT ALL:

- Battery charge low
- Electrical connections loose, dirty or defective
- Engine in gear
- Oil of wrong grade
- Starter faulty
- Starting procedure incorrect

THE ENGINE TURNS OVER BUT WILL NOT START:

- Air in fuel system
- Air filter blocked
- Cold start aid (if fitted) faulty
- Compression poor
- Engine in gear
- Fuel contaminated by water (or ice)
- Fuel filter blocked
- Fuel lift pump faulty (or sucking air through leak on suction side)
- Fuel of wrong grade or quality
- Fuel pipes blocked or leaking
- Fuel tank empty (or nearly so)
- Fuel tap shut
- Fuel 'waxing' (caused by using 'summer' fuel in low temperatures)
- Injection pump drive broken
- Injection pump faulty
- Injection pump timing wrong
- Injector(s) faulty
- Starting procedure incorrect
- Stop control faulty

THE ENGINE IS HARD TO START:

◆ Air in fuel system

◆ Air filter blocked

◆ Cold start aid (if fitted) faulty

◆ Compression poor (see page 83)

◆ Exhaust pipe blocked

◆ Fuel contaminated by water (or ice)

◆ Fuel filter blocked

◆ Fuel lift pump faulty (or sucking air through leak on suction side)

◆ Fuel tank vent blocked

◆ Fuel of wrong grade or quality

◆ Fuel pipes blocked or leaking

◆ Fuel tap shut

◆ Fuel 'waxing' (caused by using 'summer' fuel in low temperatures)

◆ Injection pump faulty

◆ Injection pump timing wrong

◆ Injector(s) faulty

◆ Starting procedure incorrect

◆ Stop control faulty

PROBLEMS SHOWN UP BY THE GAUGES

THE ENGINE OVERHEATS:

◆ Air filter blocked

◆ Cold start aid (if fitted) faulty

◆ Cooling water low

◆ Cylinder head gasket faulty

- Exhaust pipe blocked
- Gauge faulty
- Heat exchanger blocked
- Injection pump faulty
- Injection pump timing wrong
- Injector(s) faulty
- Raw-water filter blocked
- Raw-water pump faulty (eg impeller worn or broken)
- Raw-water system blocked or seacock shut
- Thermostat faulty
- Valve timing incorrect
- Valves leaking
- Vee-belt on water pump broken, loose or greasy

THE OIL PRESSURE IS UNUSUALLY LOW:

- Bearings worn
- Gauge faulty
- Oil cooler clogged
- Oil filter clogged
- Oil level low
- Oil of wrong grade
- Oil pressure valve faulty
- Oil pump faulty

THE OIL PRESSURE IS UNUSUALLY HIGH:

- Gauge faulty
- Oil of wrong grade
- Oil pressure valve faulty

SMOKE

THE ENGINE PRODUCES BLACK OR GREY EXHAUST SMOKE:

♦ Aftercooler clogged
♦ Air filter blocked
♦ Cold start aid (if fitted) faulty
♦ Compression poor (see page 83)
♦ Excessive load caused by dirty hull or too much weight on board
♦ Excessive power losses due to internal friction in engine or transmission
♦ Exhaust pipe blocked
♦ Fuel of wrong grade or quality
♦ Injection pump faulty
♦ Injection pump timing wrong
♦ Injector(s) faulty
♦ Inlet manifold leaking
♦ Leak-off pipe blocked
♦ Propeller damaged or wrong size
♦ Thermostat faulty
♦ Turbocharger faulty
♦ Ventilation to engine room inadequate

THE ENGINE PRODUCES BLUE OR WHITE EXHAUST SMOKE:

♦ Air filter blocked
♦ Cold start aid (if fitted) faulty
♦ Crankcase breathers clogged
♦ Compression poor (see page 83)
♦ Oil level too high
♦ Oil of wrong grade
♦ Thermostat faulty
♦ Turbocharger oil seals worn

UNUSUAL NOISES OR BEHAVIOUR

THE ENGINE MISFIRES:

♦ Air filter blocked
♦ Air in fuel system
♦ Cold start aid (if fitted) faulty
♦ Compression poor (see page 83)
♦ Fuel filter blocked
♦ Fuel lift pump faulty (or sucking air through leak on suction side)
♦ Fuel pipes blocked or leaking
♦ Injection pump faulty
♦ Injection pump timing wrong
♦ Injector pipe(s) distorted or broken
♦ Injector pipe(s) leaking
♦ Injector(s) faulty
♦ Overheating

ENGINE IS UNUSUALLY NOISY, WITH A HAMMERING OR KNOCKING SOUND:

♦ Bearings worn
♦ Cold start aid (if fitted) faulty
♦ Engine mountings loose or broken
♦ Fuel lift pump faulty (or sucking air through leak on suction side)
♦ Fuel of wrong grade or quality
♦ Injector(s) faulty
♦ Oil level low
♦ Overheating
♦ Piston rings or cylinder bores worn
♦ Shaft coupling loose
♦ Valve clearances incorrect
♦ Valve spring broken

- Valve timing incorrect
- Valves sticking

ENGINE RUNS UNEVENLY, 'HUNTS' OR SURGES:

- Air filter blocked
- Air in fuel system
- Cold start aid (if fitted) faulty
- Control system faulty
- Compression poor (see page 83)
- Fuel filter blocked
- Fuel lift pump faulty (or sucking air through leak on suction side)
- Fuel pipes blocked or leaking
- Fuel tank empty (or nearly so)
- Fuel tank vent blocked
- Injection pump faulty
- Injector pipe(s) distorted or broken
- Injector(s) faulty
- Stop control faulty

THE ENGINE VIBRATES EXCESSIVELY:

- Air filter blocked
- Control system faulty
- Compression poor (see page 83)
- Engine and shaft out of alignment
- Engine mountings loose or broken
- Fuel of wrong grade or quality
- Injection pump faulty
- Injector pipe(s) distorted or broken
- Injector(s) faulty
- Propeller damaged or wrong size

- Propeller or shaft bent, broken or unbalanced
- Shaft coupling loose
- Shaft misaligned

THE ENGINE LACKS POWER:

- Air in fuel system
- Air filter blocked
- Control system faulty
- Compression poor (see page 83)
- Excessive load caused by dirty hull or too much weight on board
- Excessive power losses due to internal friction in engine or transmission
- Exhaust pipe blocked
- Fuel filter blocked
- Fuel lift pump faulty (or sucking air through leak on suction side)
- Fuel of wrong grade or quality
- Fuel pipes blocked or leaking
- Fuel tank vent blocked
- Fuel 'waxing' (caused by using 'summer' fuel in low temperatures)
- Injection pump faulty
- Injection pump timing wrong
- Injector(s) faulty
- Inlet manifold leaking
- Overheating
- Propeller damaged or wrong size
- Thermostat faulty
- Turbocharger faulty
- Ventilation to engine compartment inadequate

THE ENGINE USES MORE FUEL THAN USUAL:

- Air filter blocked
- Cold start aid (if fitted) faulty

- Control system faulty
- Compression poor (see below)
- Excessive load caused by dirty hull or too much weight on board
- Excessive power losses due to internal friction in engine or transmission
- Exhaust pipe blocked
- Fuel of wrong grade or quality
- Injection pump faulty
- Injection pump timing wrong
- Injector(s) faulty
- Inlet manifold leaking
- Leak-off pipe blocked
- Thermostat faulty
- Turbocharger faulty

COMPRESSION POOR

- Cylinder head gasket faulty
- Piston rings or cylinder bores worn
- Valve clearances incorrect
- Valve guides in cylinder head worn
- Valve spring broken
- Valve timing incorrect
- Valves leaking
- Valves sticking

WINTERISING

13

A few months of idleness over the winter can do far more harm to an engine than any amount of hard work.

Oil that has done its cleaning job throughout the season lies in the engine, contaminated by the acidic by-products of combustion, while the upper part of the engine, now dry of oil, is exposed to the damp and possibly salty atmosphere. Things that are supposed to be flexible, such as vee-belts and pump impellers, may become stiff; mice and spiders hibernate in air filters and exhaust pipes; and if you're unlucky, the air temperature may fall low enough to freeze the water in the cooling system, to crack the engine block or cylinder head.

Winterisation, then, is primarily about protecting the engine against corrosion and cold. It's also an opportunity to do a lot of routine maintenance jobs.

It's important to plan what you're going to do, and when and how you're going to do it: some jobs are best done afloat; others can only be done with the boat ashore, and although some can be left until spring, others should really be done in the autumn. The exact process will vary between different boats and circumstances, but the following routine is fairly typical:

AUTUMN: *BEFORE LIFTING OUT*

♦ If possible, fill the fuel tank to reduce the air space in which condensation can form (but check that this won't take the boat over the weight limit for the crane or boat-lift).

♦ Drain the fresh-water cooling system. It's often easiest to do this by disconnecting a hose from the circulating pump. Then shut any taps or drains or reconnect any hoses, and refill with a strong solution of fresh antifreeze.

♦ Drain off any water and sludge from the fuel pre-filter, and change the fuel filter(s).

♦ Run the engine to operating temperature.

♦ Change the oil and oil filter while the engine is warm.

AUTUMN: AFTER LIFTING OUT

♦ Clean the raw-water filter.

♦ Flush the raw-water system. There are various methods, but this is one of the most widely recommended:

a. Remove the thermostat.

b. Disconnect the raw-water inlet hose from the skin fitting or seacock, and extend it with an additional length of hose if necessary so that the end can be immersed in a bucket of water. Support the bucket at or near the boat's normal waterline, and arrange for it to be kept topped up by a constant supply of fresh water from a hose.

c. Start the engine and let it run for a few minutes to flush the system. Tip antifreeze solution into the bucket, turn off the hose, and switch off the engine just before it empties the bucket.

d. Re-connect the hose to the skin fitting.

♦ Remove the raw-water pump impeller. (Put it in a plastic bag and hang it on the engine controls or tie it to the keys so that you don't start the engine without it!)

♦ Remove the alternator and water pump drive belts.

♦ Change the air filter, and seal the air intake with oiled rags and/or a plastic bag.

♦ Seal the exhaust pipe with oiled rags or a plastic bag.

♦ Inspect the gearbox/saildrive/outdrive oil, and replace it if it looks grey or milky.

♦ Remove the propeller and grease the shaft (not necessary for shaft-drive boats).

♦ Either remove the batteries to storage ashore or clean and grease their terminals and arrange for them to be charged once every six weeks or so. **All lead-acid batteries lose their charge over time, and a flat battery can then be irreparably damaged by freezing.**

♦ Spray the backs of switchboards, instrument panels and other electrical connections with water-repellent oil.

SPRING: BEFORE LAUNCHING

◆ For saildrives: inspect the rubber seal around the drive leg, and replace in accordance with the manufacturer's instructions (eg every five to seven years).

◆ For outdrives: inspect the bellows between the drive leg and transom shield, and replace in accordance with the manufacturer's instructions (eg every two years).

◆ Test and re-fit the thermostat.

◆ If your raw-water cooling system includes zinc anodes, inspect them and replace them if necessary.

◆ Check that all plugs and drain cocks in the cooling system are shut.

◆ Re-fit the raw-water pump impeller with a new pump gasket.

◆ Inspect all hoses: replace any that are visibly cracked or that do not feel pliable.

◆ Check hose clamps for security: tighten any that are loose, and replace any that are corroded.

◆ Re-fit the alternator and water pump drive belts, and adjust them.

◆ Unblock the air filter and exhaust.

◆ Check the oil level in the gearbox/saildrive/outdrive.

◆ Grease the propeller shaft and re-fit the propeller.

◆ Reconnect the batteries.

SPRING: AFTER LAUNCHING

◆ Inspect for leaks.

◆ Check the engine oil level.

◆ Grease or bleed the stern gland as appropriate.

◆ Run the engine to operating temperature, inspect for leaks, and check oil and water levels.

APPENDIX
THE RYA DIESEL ENGINE COURSE

AIM OF THE COURSE

To give an awareness of the main systems of a marine diesel engine so that the yachtsman can take simple measures to prevent mechanical breakdown at sea and rectify defects which do not require workshop support. No pre-course knowledge is required.

COURSE SYLLABUS

1 INTRODUCTION

Principles of the diesel engine

2 THE FOUR-STROKE CYCLE

- Naturally aspirated engines
- Turbocharging
- Intercooling/aftercooling

3 THE FUEL SYSTEM

- The basic system
- The tank
- The water separating pre-filter
- Fuel lift pump
- The engine fine filter
- Injection pump
- Injectors
- Bleeding the system

4 THE COOLING SYSTEM

- Seawater cooling
- Freshwater cooling
- Temperature control
- The thermostat
- The seawater impeller pump

5 THE AIR SYSTEMS

- The airway in
- The airway out

6 ENGINE ELECTRICAL SYSTEMS

- The basic system
- Battery capacity and care
- Drive belts
- The alternator

7 SPARES AND TOOL REQUIREMENTS

- Basic spares and tools

8 IMPORTANCE OF WINTERISATION AND SERVICING

- Engine lubrication
- Transmission lubrication
- Winterisation and servicing
- Service schedule
- Winterisation

9 FAULT-FINDING

COURSE REQUIREMENTS

The minimum duration of the course is six hours. A diesel engine (not necessarily in working condition) will be provided for practical sessions. (No more that six students to one engine.) Instructors will have attended an RYA Training Course.

Enquiries to: Royal Yachting Association, RYA House, Ensign Way, Hamble, Southampton SO31 4YA. Tel: 023 8060 4100 Website: www.rya.org.uk

INDEX